Advanced Research in Age-Related Macular Degeneration (AMD)

Advanced Research in Age-Related Macular Degeneration (AMD)

Editor

Oyuna Kozhevnikova

MDPI

Basel • Beijing • Wuhan • Barcelona • Belgrade • Novi Sad • Cluj • Manchester

Editor
Oyuna Kozhevnikova
Institute of Cytology and
Genetics, Siberian Branch of
Russian Academy of Sciences
Novosibirsk
Russia

Editorial Office
MDPI
St. Alban-Anlage 66
4052 Basel, Switzerland

This is a reprint of articles from the Special Issue published online in the open access journal *Biomedicines* (ISSN 2227-9059) (available at: https://www.mdpi.com/journal/biomedicines/special_issues/Age_Related_Macular_Degeneration).

For citation purposes, cite each article independently as indicated on the article page online and as indicated below:

Lastname, A.A.; Lastname, B.B. Article Title. *Journal Name* **Year**, *Volume Number*, Page Range.

ISBN 978-3-7258-1177-9 (Hbk)
ISBN 978-3-7258-1178-6 (PDF)
doi.org/10.3390/books978-3-7258-1178-6

Contents

About the Editor

Oyuna Kozhevnikova

Oyuna Kozhevnikova is the leading researcher at the Institute of Cytology and Genetics SB RAS, Novosibirsk, Russia. After graduating with an MSc degree in Biology in 2007 from Novosibirsk State University, they pursued a PhD in Genetics in 2014. Their dissertation was entitled "Transcriptome changes in the retina of rats with age in the development of retinopathy". During their PhD training, they contributed to the characterization of a unique senescent-accelerated rat model of age-related diseases. They continued to develop this stream of research during their postdoctoral training at the Institute of Cytology and Genetics (Prof Kolosova's lab), where they participated in a number of projects focusing on retinal degeneration, including the organization of studies of the molecular genetic mechanisms of retinal aging and the development of age-dependent eye diseases. Their current main research interests are the molecular mechanisms of retinal degeneration, the population genetics of neovascular age-related macular degeneration, the role of autophagy in age-related macular degeneration, microRNAs as biomarkers of poor response to antiangiogenic treatment, biomarkers and early predictors of age-related diseases, transcriptomics, cellular aging, and the search for pharmacological targets for age-related diseases.

biomedicines

Editorial

Advanced Research in Age-Related Macular Degeneration: Special Issue

Oyuna Kozhevnikova

Federal Research Center Institute of Cytology and Genetics SB RAS, Pr. Lavrentiev, 10, 630090 Novosibirsk, Russia; oidopova@bionet.nsc.ru

Age-related macular degeneration (AMD) is an eye disease that is the leading cause of irreversible vision loss in people over 55 years of age [1]. Due to the aging of the global population predicted in the coming decades, the number of patients with AMD is expected to increase significantly [2]. AMD is a disease with a complex nature, the development of which is controlled by many interacting genetic and environmental factors that determine the form, rate of disease progression, and response to therapy [3]. The pathogenesis of AMD is based on a decrease in metabolic and regenerative processes characteristic of aging, a violation of the microcirculation and structural organization of the retina, the prerequisites and mechanisms for the transition of which into a pathological process remain unclear [4]. Therefore, studies of the molecular mechanisms of AMD pathogenesis and new approaches to its early diagnosis, treatment, and prevention are urgently needed.

Late-stage AMD is distinguished by the neovascularization (wet type) or geographic atrophy (GA) of the retinal pigment epithelium (RPE) cell layer (dry type). There is growing genetic, experimental, and clinical evidence supporting an association between the pathophysiology of dry AMD and key proteins in the complement cascade. As a fundamental component of the innate immune system, the complement cascade protects the body from foreign pathogens and altered self-tissues. This highlights the crucial role that inflammation plays in the onset and progression of the disease. In a review, Patel et al. [5] provide an understanding of the role of the complement system in dry AMD and discuss emerging therapies in early-phase clinical trials. It is hoped that these drugs could potentially intervene early in the pathogenesis of dry AMD, halting the progression of the disease to a later stage. This pathway provides an abundant target for the synthesis of novel therapeutics. One of them—pegcetacoplan, additionally denoted as APL-2—is a PEGylated peptide inhibitor of C3 formulated by Apellis Pharmaceuticals that is administered intravitreally [6]. Moreover, on 17 February 2023, the Food and Drug Administration approved Syfovre® (intravitreal pegcetacoplan, 15 mg, Apellis Pharmaceuticals, Waltham, MA, USA) based on a slower progression of atrophy in the DERBY and OAKS phase 3 randomized controlled clinical trials [7].

In a tutorial, Biarnés et al. [7] critically appraise the methodological aspects of phase 3 clinical trials in GA in relation to their design, analysis, and interpretation. The authors reviewed some of the key methodological attributes of phase 3 clinical trials in GA available in the main public registry of clinical trials as of 20 May 2023 (GATHER1, DERBY/OAKS, CHROMA/SPECTRI, SEATTLE, and GATE) to improve their interpretation. The topics discussed included types of endpoints, eligibility criteria, p-value and effect size, study power and sample size, the intention to treat principle, missing data, the consistency of results, efficacy–safety balance, and the application of results [7]. While these clinical studies share many outcomes and general eligibility requirements, there are several aspects that might affect outcomes which are discussed in this work.

The quantitative autofluorescence (qAF8) level is a presumed surrogate marker of lipofuscin content in the retina [8]. Chandra et al. [9] investigated changes in qAF8 levels in non-neovascular AMD in a prospective cohort study. The AMD categories were graded

Citation: Kozhevnikova, O. Advanced Research in Age-Related Macular Degeneration: Special Issue. *Biomedicines* **2024**, *12*, 392. https://doi.org/10.3390/biomedicines12020392

Received: 15 January 2024
Accepted: 30 January 2024
Published: 8 February 2024

using both the Beckman classification and multimodal imaging, allowing for the deep phenotyping of AMD. A total of 353 eyes were analyzed, representing the largest study of qAF8 in AMD to date. The results showed that qAF8 is decreased during the early progression of AMD. qAF8 levels seem related more to the loss of function and integrity of RPE cells rather than being due to abnormalities in the visual cycle [9].

Zhdankina et al. [10] assessed the retinoprotective potential of the JNK inhibitor 11H-indeno[1,2-b]quinoxalin-11-one oxime sodium salt (IQ-1S), using senescence-accelerated OXYS rats as a model of AMD. Retinopathy in the OXYS rats reproduced the major signs of dry AMD: dystrophic alterations to the RPE, thinning of the neuroretina, and impaired choroidal microcirculation [11]. Treatment with IQ-1S significantly improved the structural and functional parameters of RPE. This suggests that an increase in the JNK-signaling pathway may play a role in the treatment of dry AMD.

AMD shares some risk factors with diabetes mellitus, but little is known about the risk of DM in individuals with AMD. Jung et al. [12] investigated the association between AMD and the risk of diabetes mellitus using the Korean Nationwide Health Insurance Database. A Cox hazard regression model was used to examine the hazard ratios of diabetes mellitus with adjustments for potential confounders. The study did not find an increased risk of diabetes mellitus in individuals with AMD.

With the availability of high-resolution retinal imaging, the classification of macular neovascularization (MNV) and drusen has evolved. Optical coherence tomography (OCT) has improved our understanding of MNV types and has expanded and refined distinctions between drusen types. Subretinal drusenoid deposits (SDDs) have a high prevalence in AMD which was underestimated prior to OCT [13]. The purpose of the study by Muth et al. [14] was to assess the predictive odds ratios between drusen types and MNV lesion variants in patients with treatment-naïve neovascular AMD. Multimodal imaging was retrospectively reviewed for drusen type (soft drusen, SDD, or mixed) and MNV type according to CONAN criteria [15]. The authors concluded that SDDs represent a common phenotypic characteristic in AMD eyes with treatment-naïve MNV. The adjusted odds ratio for eyes with predominant SDDs to develop MNV 2 and MNV 3 was much higher, possibly due to their location in the subretinal space.

One study [16] aimed to report the clinical and multimodal imaging findings of patients with neovascular AMD who developed bacillary layer detachment (BALAD). This finding is not commonly described in the literature. BALAD is a definition introduced in 2018 to describe an OCT finding in a patient with Toxoplasmosis chorioretinitis and pachychoroid disease [17]. It is characterized by a splitting of the photoreceptor layer at the myoid level, resulting in a space posterior to the external limiting membrane.

Acquired vitelliform lesions (AVLs) are associated with a large spectrum of retinal diseases, including AMD [18]. The purpose of the study by Damian et al. [19] was to characterize AVLs in AMD patients using OCT and ImageJ software (version 1.53T). A vitelliform lesion represents an accumulation of lipofuscin, melanosomes, melanolipofuscin, and outer segment debris in the subretinal space [20]. The accumulation of subretinal fluid associated with vitelliform lesions was explained by a mechanical displacement between the RPE and outer retinal layers, which consequently inhibits the RPE from pumping out liquified lipofuscin debris [18,19]. The authors found increased RPE thickness as a sign of hyperplasia, contrary to the ONL layer, which was decreased, mirroring the impact of the vitelliform lesion over the photoreceptors. Most of the analyzed eyes presented a discontinuous ellipsoid zone. During follow-up, some of the AVLs had increased while others had regressed, which was demonstrated in divergent results between eyes regarding height, width, area, perimeter, or volume [19].

Neovascular AMD patients with alternative alleles of the *STAT4* genes rs10181656 and rs10168266 exhibited significantly lower serum STAT4 levels than the control group, suggesting a link between specific *STAT4* genotypes and serum STAT4 levels in AMD patients [21].

The first-line drugs for neovascular AMD are inhibitors of VEGF, with up to 30% of patients having an incomplete response to treatment [22]. Genetic factors may influence the response to anti-VEGF therapy and explain treatment outcome variability. Recently, autophagy was implicated in causing tumor resistance to antiangiogenic therapy [23], which suggests an analogous connection between autophagy and anti-VEGF intravitreal injections during AMD treatment. In this direction, a study was conducted assessing the role of polymorphic markers of autophagy genes in the risk of AMD and the anti-VEGF response [24]. It was shown that *MTOR* gene polymorphisms were moderately associated with neovascular AMD. The variants *SQSTM1*-rs10277 and *ULK1*-rs3088051 may influence the short-term response to anti-VEGF treatment. The results suggest that autophagy could be a target for future drugs to overcome resistance to anti-VEGF therapy [24].

Through published articles, this Special Issue provides new insight into the diagnosis and treatment of AMD and has the potential to impact current and future knowledge of the pathogenesis and treatment of patients with AMD. New diagnostic approaches and treatment regimens, the development of multimodal visualization methods and potential genetic and OCT markers of treatment response may help guide the future management of this disease.

Funding: This work was supported by the State Budget Project FWNR-2022-0016.

Acknowledgments: I am grateful to the authors of the published papers for their valuable contributions and to the reviewers for the rigorous reviews carried out. I express my gratitude to the editorial board of the journal.

Conflicts of Interest: The author declares no conflict of interest.

References

1. Guymer, R.H.; Campbell, T.G. Age-related macular degeneration. *Lancet* **2023**, *401*, 1459–1472. [CrossRef]
2. Mitchell, P.; Liew, G.; Gopinath, B.; Wong, T.Y. Age-related macular degeneration. *Lancet* **2018**, *392*, 1147–1159. [CrossRef]
3. Fursova, A.Z.; Derbeneva, A.S.; Vasilyeva, M.A.; Nikulich, I.F.; Tarasov, M.S.; Gamza, Y.A.; Chubar, N.V.; Gusarevitch, O.G.; Dmitrieva, E.I.; Kozhevnikova, O.S.; et al. New findings on pathogenetic mechanisms in the development of age-related macular degeneration. *Vestn. Oftalmol.* **2022**, *138*, 120–130. [CrossRef] [PubMed]
4. Telegina, D.V.; Kozhevnikova, O.S.; Kolosova, N.G. Molecular mechanisms of cell death in retina during development of age-related macular degeneration. *Adv. Gerontol.* **2017**, *7*, 17–24. [CrossRef]
5. Patel, P.N.; Patel, P.A.; Land, M.R.; Bakerkhatib-Taha, I.; Ahmed, H.; Sheth, V. Targeting the Complement Cascade for Treatment of Dry Age-Related Macular Degeneration. *Biomedicines* **2022**, *10*, 1884. [CrossRef]
6. Liao, D.S.; Grossi, F.V.; El Mehdi, D.; Gerber, M.R.; Brown, D.M.; Heier, J.S.; Wykoff, C.C.; Singerman, L.J.; Abraham, P.; Grassmann, F.; et al. Complement C3 Inhibitor Pegcetacoplan for Geographic Atrophy Secondary to Age-Related Macular Degeneration: A Randomized Phase 2 Trial. *Ophthalmology* **2020**, *127*, 186–195. [CrossRef] [PubMed]
7. Biarnés, M.; Garrell-Salat, X.; Gómez-Benlloch, A.; Guarro, M.; Londoño, G.; López, E.; Ruiz, S.; Vázquez, M.; Sararols, L. Methodological Appraisal of Phase 3 Clinical Trials in Geographic Atrophy. *Biomedicines* **2023**, *11*, 1548. [CrossRef] [PubMed]
8. Sparrow, J.R.; Duncker, T.; Schuerch, K.; Paavo, M.; de Carvalho, J.R.L., Jr. Lessons learned from quantitative fundus autofluorescence. *Prog. Retin. Eye Res.* **2020**, *74*, 100774. [CrossRef] [PubMed]
9. Chandra, S.; Grewal, M.K.; Gurudas, S.; Sondh, R.; Bird, A.; Jeffery, G.; Chong, V.; Sivaprasad, S. Quantitative Autofluorescence in Non-Neovascular Age Related Macular Degeneration. *Biomedicines* **2023**, *11*, 560. [CrossRef] [PubMed]
10. Zhdankina, A.A.; Tikhonov, D.I.; Logvinov, S.V.; Plotnikov, M.B.; Khlebnikov, A.I.; Kolosova, N.G. Suppression of Age-Related Macular Degeneration-like Pathology by c-Jun N-Terminal Kinase Inhibitor IQ-1S. *Biomedicines* **2023**, *11*, 395. [CrossRef]
11. Kozhevnikova, O.S.; Telegina, D.V.; Tyumentsev, M.A.; Kolosova, N.G. Disruptions of Autophagy in the Rat Retina with Age During the Development of Age-Related-Macular-Degeneration-like Retinopathy. *Int. J. Mol. Sci.* **2019**, *20*, 4804. [CrossRef]
12. Jung, W.; Yoon, J.M.; Han, K.; Kim, B.; Hwang, S.; Lim, D.H.; Shin, D.W. Association between Age-Related Macular Degeneration and the Risk of Diabetes Mellitus: A Nationwide Cohort Study. *Biomedicines* **2022**, *10*, 2435. [CrossRef]
13. Zweifel, S.A.; Spaide, R.F.; Curcio, C.A.; Malek, G.; Imamura, Y. Reticular pseudodrusen are subretinal drusenoid deposits. *Ophthalmology* **2010**, *117*, 303–312.e1. [CrossRef]
14. Muth, D.R.; Toro, M.D.; Bajka, A.; Jonak, K.; Rieder, R.; Kohler, M.M.; Gunzinger, J.M.; Souied, E.H.; Engelbert, M.; Freund, K.B.; et al. Correlation between Macular Neovascularization (MNV) Type and Druse Type in Neovascular Age-Related Macular Degeneration (AMD) Based on the CONAN Classification. *Biomedicines* **2022**, *10*, 2370. [CrossRef]

15. Spaide, R.F.; Jaffe, G.J.; Sarraf, D.; Freund, K.B.; Sadda, S.R.; Staurenghi, G.; Waheed, N.K.; Chakravarthy, U.; Rosenfeld, P.J.; Holz, F.G.; et al. Consensus Nomenclature for Reporting Neovascular Age-Related Macular Degeneration Data: Consensus on Neovascular Age-Related Macular Degeneration Nomenclature Study Group. *Ophthalmology* **2020**, *127*, 616–636. [CrossRef]
16. Palmieri, F.; Younis, S.; Raslan, W.; Fabozzi, L. Bacillary Layer Detachment in Neovascular Age-Related Macular Degeneration: Case Series. *Biomedicines* **2023**, *11*, 988. [CrossRef]
17. Mehta, N.; Chong, J.; Tsui, E.; Duncan, J.L.; Curcio, C.A.; Freund, K.B.; Modi, Y. Presumed foveal bacillary layer detachment in a patient with toxoplasmosis chorioretinitis and pachychoroid disease. *Retin. Cases Brief Rep.* **2021**, *15*, 391–398. [CrossRef]
18. Juliano, J.; Patel, S.; Ameri, H. Acquired Vitelliform Macular Degeneration: Characteristics and Challenges of Managing Subretinal Fluid. *J. Ophthalmic Vis. Res.* **2021**, *16*, 582–591. [CrossRef]
19. Damian, I.; Muntean, G.-A.; Galea-Holhoș, L.-B.; Nicoară, S.-D. Advanced ImageJ Analysis in Degenerative Acquired Vitelliform Lesions Using Techniques Based on Optical Coherence Tomography. *Biomedicines* **2023**, *11*, 1382. [CrossRef]
20. Balaratnasingam, C.; Hoang, Q.V.; Inoue, M.; Curcio, C.A.; Dolz-Marco, R.; Yannuzzi, N.A.; Dhrami-Gavazi, E.; Yannuzzi, L.A.; Freund, K.B. Clinical Characteristics, Choroidal Neovascularization, and Predictors of Visual Outcomes in Acquired Vitelliform Lesions. *Am. J. Ophthalmol.* **2016**, *172*, 28–38. [CrossRef]
21. Blekeris, T.; Gedvilaite, G.; Kaikaryte, K.; Kriauciuniene, L.; Zaliuniene, D.; Liutkevciene, R. Association of STAT4 Gene Polymorphisms (rs10181656, rs7574865, rs7601754, rs10168266) and Serum STAT4 Levels in Age-Related Macular Degeneration. *Biomedicines* **2024**, *12*, 18. [CrossRef] [PubMed]
22. Yang, S.; Zhao, J.; Sun, X. Resistance to anti-VEGF therapy in neovascular age-related macular degeneration: A comprehensive review. *Drug Des. Dev. Ther.* **2016**, *10*, 1857–1867.
23. Chandra, A.; Rick, J.; Yagnik, G.; Aghi, M.K. Autophagy as a mechanism for anti-angiogenic therapy resistance. *Semin. Cancer Biol.* **2020**, *66*, 75–88. [CrossRef] [PubMed]
24. Kozhevnikova, O.S.; Fursova, A.Z.; Derbeneva, A.S.; Nikulich, I.F.; Devyatkin, V.A.; Kolosova, N.G. Pharmacogenetic Association between Allelic Variants of the Autophagy-Related Genes and Anti-Vascular Endothelial Growth Factor Treatment Response in Neovascular Age-Related Macular Degeneration. *Biomedicines* **2023**, *11*, 3079. [CrossRef]

biomedicines

Review

Targeting the Complement Cascade for Treatment of Dry Age-Related Macular Degeneration

Prem N. Patel [1,†], Parth A. Patel [2,†], Matthew R. Land [2], Ibrahim Bakerkhatib-Taha [3], Harris Ahmed [4] and Veeral Sheth [3,*]

1 Department of Ophthalmology, University of Texas Southwestern Medical Center, Dallas, TX 75390, USA
2 Department of Ophthalmology, Medical College of Georgia, Augusta University, Augusta, GA 30912, USA
3 University Retina and Macula Associates, Oak Forest, IL 60452, USA
4 Department of Ophthalmology, Loma Linda University Medical Center, Loma Linda, CA 92350, USA
* Correspondence: vsheth@uretina.com
† These authors contributed equally to this work.

Citation: Patel, P.N.; Patel, P.A.; Land, M.R.; Bakerkhatib-Taha, I.; Ahmed, H.; Sheth, V. Targeting the Complement Cascade for Treatment of Dry Age-Related Macular Degeneration. *Biomedicines* **2022**, *10*, 1884. https://doi.org/10.3390/biomedicines10081884

Academic Editor: Oyuna Kozhevnikova

Received: 30 June 2022
Accepted: 1 August 2022
Published: 4 August 2022

Abstract: Age-related macular degeneration (AMD) is the leading cause of irreversible vision loss in the elderly population. AMD is characterized in its late form by neovascularization (wet type) or geographic atrophy of the retinal pigment epithelium cell layer (dry type). Regarding the latter type, there is growing evidence supporting an association between the pathophysiology of dry AMD and key proteins in the complement cascade. The complement cascade works as a central part of the innate immune system by defending against foreign pathogens and modified self-tissues. Through three distinct pathways, a series of plasma and membrane-associated serum proteins are activated upon identification of a foreign entity. Several of these proteins have been implicated in the development and progression of dry AMD. Potential therapeutic targets include C1q, C3, C5, complement factors (B, D, H, I), membrane attack complex, and properdin. In this review, we provide an understanding of the role of the complement system in dry AMD and discuss the emerging therapies in early phase clinical trials. The tentative hope is that these drugs may offer the potential to intervene at earlier stages in dry AMD pathogenesis, thereby preventing progression to late disease.

Keywords: age-related macular degeneration; geographic atrophy; complement cascade; therapeutics; clinical trials

1. Introduction

Age-related macular degeneration (AMD) is a debilitating condition that represents one of the leading causes of visual impairment among the elderly population in developed countries [1]. Estimates currently place the global affected population between 30 and 50 million individuals, a figure expected to rise to almost 300 million by 2040 [2]. As a consequence of its prevalence and the associated progressive and inexorable central vision loss, the societal burden of this disease is immense [3].

Classification of AMD is established on clinical observation. Although multiple systems have been developed [4–6], the predominant scale defines disease progression into three stages according to features such as drusen size and absence or presence of pigmentary changes, geographic atrophy (GA), and neovascularization. Early and intermediate stage disease is characterized by variation in the presence and size of drusen (>63 μm) and potential retinal pigment epithelium (RPE) abnormalities. Late stage disease is characterized by elements of substantial advancement, including GA and choroidal neovascularization (CNV) [7]. GA is the definitive outcome in the progression of nonexudative (dry) AMD, whereby degeneration of macular photoreceptors and RPE cells irreversibly atrophies retinal tissue. Alternatively, CNV is the hallmark of exudative (wet) AMD, in which abnormal blood vessels proliferate underneath the retina and macula, with resultant leakage, hemorrhage, and possible pigment epithelial detachment.

Relative to its dry counterpart, wet AMD has traditionally resulted in a greater proportion of patients with vision loss; however, the discovery of the association of vascular endothelial growth factor (VEGF) with its pathophysiology and subsequent development of targeted therapeutics has been a boon for its treatment. While anti-VEGF medications (e.g., aflibercept, brolucizumab, ranibizumab, etc.) remain imperfect and fail to demonstrate a clinical response in certain patients [8], their positive effects on long-term visual outcomes are well-established [9–11]. Unfortunately, treatment for dry AMD remains elusive, partially due to an incomplete understanding of the mechanisms involved in its pathogenesis.

Dysregulation in the complement cascade exhibits a particularly robust relationship with dry AMD [12–14], offering a strategy for treatments specific to the disease. Considering the constant advances occurring in our understanding of GA, the objective of this review is to provide a discussion of the broad mechanisms implicated in the progression of dry AMD while systematically focusing on the therapeutics previously and currently in development to target the complement cascade (Table 1).

Table 1. Completed and current investigational treatments targeting the complement cascade in dry age-related macular degeneration.

Therapeutic Category or Name	Mechanism	Mode of Administration	Clinical Trial ID, NCT# *	Study Phase (Status) †	Patients, *n*	Primary Outcome
C1Q						
ANX007	Recombinant monoclonal antibody	Intravitreal	NCT04656561	Phase II (Recruiting)	240 (Anticipated)	GA growth
C3						
AMY-106 (Cp40-KKK)	Compstatin analog inhibitor	Intravitreal	N/A	N/A	N/A	N/A
Pegcetacoplan (APL-2)	PEGylated peptide inhibitor	Intravitreal	NCT03525600	Phase III (Ongoing)	621	GA growth
				Phase III (Ongoing)		
			NCT03525613	Phase III (Recruiting)	638	GA growth
			NCT04770545	Phase II (Terminat-ed, Adverse effects)	1200 (Anticipated)	Safety
				Phase II (Ongoing)		
POT-4 (AL-78898A)	Compstatin analog inhibitor	Intravitreal	NCT01603043		10	GA growth
NGM621	Humanized monoclonal antibody	Intravitreal	NCT04465955		320	GA growth
C5						
Eculizumab	Humanized monoclonal antibody	Intravenous	NCT00935883	Phase II (Completed, Ineffective)	60	GA growth/Reduction of drusen volume
						GA growth
				Phase III (Ongoing)		
	PEGylated RNA aptamer					
Avacincaptad Pegol (Zimura® [ARC1905]) Tesidolumab (LFG316)		Intravitreal	NCT04435366	Phase II (Completed, Ineffective)	448	GA growth
	Human monoclonal anti-body					
		Intravitreal	NCT01527500		158	
Complement Factor B						
IONIS-FB-LRx	Subcutaneous	NCT03815825	Phase II (Recruiting)	330 (Anticipated)	GA growth	
Complement Factor D						
Lampalizumab (FCFD4514S)	Humanized monoclonal antibody	Intravitreal	NCT02247479	Phase III (Terminated, Ineffective)	906	GA growth
				Phase III (Terminated, Ineffective)		
			NCT02247531		975	GA growth

Table 1. *Cont.*

Therapeutic Category or Name	Mechanism	Mode of Administration	Clinical Trial ID, NCT# *	Study Phase (Status) †	Patients, *n*	Primary Outcome
Complement Factor H						
AdCAGfH	Adeno-associated virus	Subretinal gene therapy	N/A	N/A	N/A	N/A
GEM103	Recombinant human CFH	Intravitreal	NCT04643886	Phase IIa (Terminated, Effective)	62	Safety
Complement Factor I						
GT005	Adeno-associated virus	Subretinal gene therapy	NCT03846193	Phase I/II (Recruiting)	65 (Anticipated)	Safety
			NCT04566445	Phase II (Recruiting)	250 (Anticipat-ed)	GA growth
			NCT04437368	Phase II (Recruiting)	75 (Anticipated)	GA growth
Membrane Attack Complex (MAC)						
AAVCAGsCD59 (HMR59)	Adeno-associated virus	Intravireal gene therapy	NCT04358471	Phase II (Hiatus)	N/A	GA growth
Properdin						
CLG561	Human antibody Fab	Intravitreal	NCT02515942	Phase II (Completed, Ineffective)	114	Safety/IOP change/GA growth

Abbreviations: NCT, National Clinical Trial; GA, geographic atrophy; IOP, intraocular pressure. * NCT# was unavailable for some therapeutics as formal clinical trials have yet not been conducted; NCT# is provided for the most recent clinical trials. † Where applicable, the most recent study phase is listed.

2. Methods of Literature Search

To identify all therapeutics targeting components of the complement cascade, a comprehensive literature search was conducted using PubMed, Google Scholar, Web of Science, Clinicaltrials.gov, and company websites. The latter enabled the identification of pertinent findings presented at recent conferences but not yet published in the literature. Articles in the English language were included exclusively. This literature search was conducted up to the end of May 2022.

3. Risk Factors

Various modifiable and nonmodifiable demographic, environmental, and genetic factors have been recognized as predicting risk for GA, although their exact roles in disease progression remain uncertain [15].

3.1. *Demographic and Environmental*

Advanced age has been established as the primary contributor to GA development, with one analysis indicating prevalence rates of late stage AMD ranging from 0.1% among individuals aged 55–59 years to 9.8% in individuals aged 85 years and older [16]. The strongest modifiable risk factor for GA is smoking, with up to a four-fold increase in risk observed [17–20]. Other modifiable factors that have been explored include body mass index (BMI), diet, physical activity, and sunlight exposure [18,19,21–26]. Unfortunately, findings from multiple studies are conflicting, and there remains a need for further research to determine conclusive associations.

3.2. *Genetic*

Genomic investigations have profoundly expanded our understanding of the influence of specific genes in the development of AMD. In the original Genome-Wide Association Study (GWAS), 52 variants among 32 genetic loci were identified, which were predominantly involved in pathways related to the complement cascade, extracellular matrix remodeling, and lipid metabolism [27]. Findings from a more recent GWAS demonstrated enrichment among the complement, immune response, and protein regulation pathways across loci associated with disease progression [28].

The discovery of polymorphisms in two primary genetic loci has specifically established the genetic predisposition of individuals towards the development of GA secondary to dry AMD. The rs1061170 variant of the complement factor H (CFH) gene increases the incidence of early AMD and eventual progression to late stage AMD [14]. Similarly, the rs10490924 variant located near the age-related maculopathy susceptibility 2 (ARMS2) and high-temperature requirement A serine peptidase 1 (HTRA1) genes are strongly attributed to elevated risk for AMD and disease progression [29,30].

Variants in genetic loci within other regions are additional considerations when assessing risk. Particularly, gene polymorphisms related to the complement cascade (C2, C3, complement factor B [CFB], and complement factor I [CFI]) and lipid metabolism (apolipoprotein E [APOE], cholesterol ester transfer protein (CETP), and lipase C [LIPC]) modulate disease incidence and progression [15,20,31–33]. Summarily stated, these variants emphasize the role of local inflammatory processes in the pathogenesis of dry AMD and its progression to GA.

4. General Pathogenesis of Dry AMD

As evidenced by prior investigations, dry AMD is a multifactorial disease entity with various genetic and environmental elements contributing toward pathogenesis that remains entirely elusive. One component of this process is the overproduction of reactive oxygen species, which translates to oxidative stress, free radical formation, and peroxidation of the RPE [34–36]. Oxidative stress particularly obstructs the functionality of RPE stem cells by inhibiting mitochondrial activity [35,37–39], thereby creating a pro-inflammatory microenvironment unconducive to appropriate retinal development. Indeed, human RPE cells exposed to greater concentrations of oxidative stress demonstrate a dose-dependent reduction in viability [40]. Compounded with other factors, these oxidative insults result in the formation of drusen, which are foundational to the progression of dry AMD [41].

This formation and enlargement of drusen inhibits the ability of Bruch's membrane (BM) to transport oxygen, nutrients, and metabolic substances by creating RPE detachments [42]. In turn, photoreceptor debris accumulates, resulting in further accretion of metabolites toxic to the RPE, such as lipofuscin and A2E [43,44]. Drusen additionally contain multiple pro-inflammatory markers, thus suggesting the importance of these processes towards the development of dry AMD [45]. These include components of the complement cascade, such as activators, fragments, and the membrane attack complex (MAC) [46]. The collective consequence of the aforementioned events is GA, a condition demarcated by outer retinal atrophic lesions that occur secondary to the loss or attenuation of the choriocapillaris, photoreceptors, and RPE [47,48].

5. Complement Cascade

The complement cascade is a component of the innate and adaptive immune systems, possessing a critical role in the defense against pathogens and the maintenance of homeostasis. Three principal pathways are involved in complement: the classical, alternative, and lectin pathways, as highlighted in Figure 1. Each is comprised of a series of reactions that ultimately conclude in the creation of a MAC [49]; nonetheless, the proteins involved in the initiation of each mechanism are distinct.

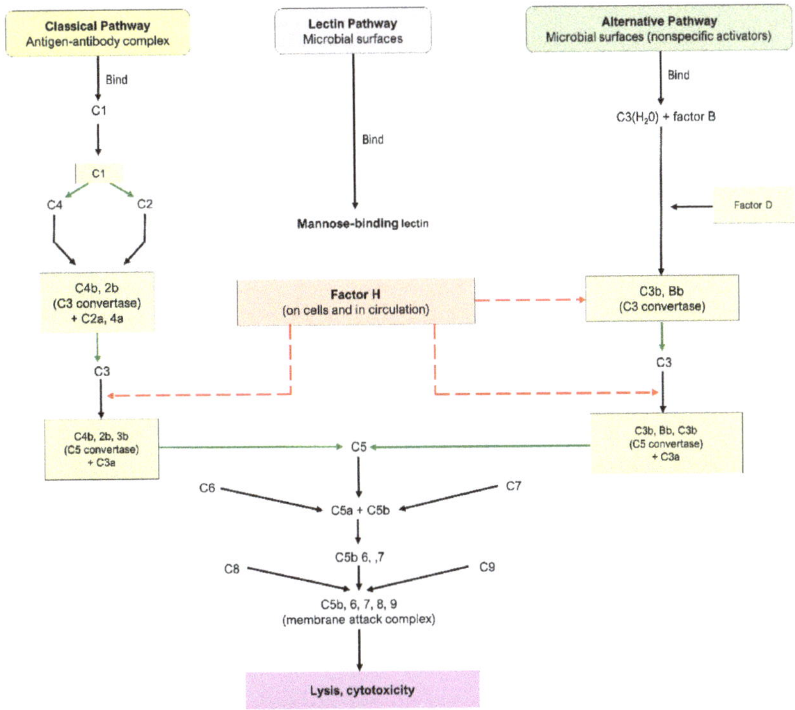

Figure 1. Complement cascade pathway.

5.1. Pathways of the Complement Cascade

The classical pathway is initiated through the binding of antigen–antibody complexes to the C1q protein, the alternative pathway is initiated through a binding interaction between the C3b protein and foreign substances, and the lectin pathway involves the interaction of mannose-binding lectin with mannose-based polysaccharides on microbials. While the lectin and alternative pathways are effector arms of innate immunity, the classical pathway is an effector arm of adaptive immunity. Nevertheless, all three pathways converge at C3 convertase, leading to the cleavage of C3 into C3a and c3b, formation of C5 convertase, cleavage of C5 into C5a and c5b, and eventual formation of the MAC and subsequent cell destruction [49]. As a consequence of their fundamental roles, C3, C5, and the MAC have emerged as crucial targets for the therapeutic pipeline of dry AMD.

5.2. Complement Cascade in AMD

As alluded to previously, overactivity of the complement pathway has been heavily implicated in the pathogenesis of drusen [50]. Several proteins involved in the complement system have been identified as possible actors in the development and progression of dry AMD, including C3, C5, CFH, CFI, complement factor D (CFD), and complement factor B (CFB).

Evidence of this association is abundant in both experimental and clinical investigations. Among mice with inactivated C3, the rate of CNV was significantly lower following laser photocoagulation, thereby highlighting the protein's role in the pathogenesis of AMD [51,52]. Other investigations have reported the presence of C5 in drusen and its elevation in the bloodstream of patients with AMD. C5a has been identified as having numerous inflammatory actions and thus is an effective target in reducing retinal loss in murine models [53,54]. MAC tends to accumulate in regions adjacent to microvascular injury, an early indicator of AMD; therefore, modulating MAC may represent a mechanism to prevent and/or limit the progression of GA [53,55,56]. Furthermore, downstream products of CFB have been observed to be elevated in AMD patients, inactivation of CFD has been linked to decreased photoreceptor loss, CFP stabilizes C3 and C5 convertases, and CD59 has been a basis for gene therapy specific to MAC [57–59].

6. Current Therapeutic Targets

Current management of dry AMD is centered around clinical observation and routine follow-up diagnostics and evaluations. Although there have been multiple clinical trials targeting components of the complement cascade for dry AMD (Table 1), an effective therapeutic is heretofore unavailable.

6.1. C1Q

ANX007

ANX007 (Annexon Inc., Brisbane, CA, USA) is a recombinant monoclonal antibody with an antigen-binding fragment that inhibits c1q. Through its action on c1q, the classical complement pathway is inhibited alongside C3 and C5 [60,61]. In a murine model of retinal photooxidative damage, ANX-M1, a monoclonal anti-C1q antibody similar to ANX007, reduced retinal atrophy [62]. Furthermore, a phase I trial (NCT04188015) using 2.5 mg and 5 mg of intravitreal (IVT) ANX007 demonstrated the safety and tolerability of both doses in 17 patients with primary open-angle glaucoma [60,61].

Because of these results, the phase II multicenter, randomized ARCHER study (NCT04656561), currently in its recruiting stage, will be conducted on an anticipated 240 patients with AMD to evaluate ANX007′s effect on GA growth rates. Patients will be randomized in a 2:2:1:1 ratio to receive ANX007 monthly, ANX007 every other month (EOM), sham monthly, or sham EOM, respectively.

6.2. C3

6.2.1. AMY-106 (Cp40-KKK)

AMY-106, derived from Cp40-KKK, a fourth-generation compstatin analog, is a novel C3 inhibitor in development by Amyndas Pharmaceuticals [63,64]. This therapeutic maintained intraocular residence for more than 90 days after one 0.5 mg IVT injection in an investigation involving cynomolgus monkeys. Moreover, AMY-106 exhibited notable retinal tissue penetrance, as it localized with C3 in the choriocapillaris [63]. Provided its promise for dry AMD treatment, a phase I trial of AMY-106 is in development [64].

6.2.2. Pegcetacoplan (APL-2)

Pegcetacoplan, additionally denoted as APL-2, is a PEGylated peptide inhibitor of C3 formulated by Apellis Pharmaceuticals that is administered intravitreally [65]. The agent was evaluated in the FILLY trial (NCT02503332) [65], a multicenter, randomized phase II trial encompassing 246 patients with GA. Subjects were randomized in a 2:2:1:1 ratio to injections of 15 mg pegcetacoplan monthly or EOM, or sham injections monthly or EOM for 12 months. Follow-up was conducted at 15 and 18 months. Pegcetacoplan treatment produced statistically significant reductions in GA growth rates. Compared to sham treatment, monthly pegcetacoplan treatment resulted in a 29% (95% confidence interval [CI], 9–49; $p = 0.008$) reduction in GA growth rate, while EOM treatment produced a 20% reduction (95% CI, 0–40; $p = 0.067$) [65].

Post hoc analysis suggested pegcetacoplan may be useful in AMD prior to the development of GA [66]. Incomplete RPE and outer retinal atrophy (iRORA) without neovascularization are considered precursor lesions to GA, while progression to complete RPE and outer retinal atrophy (cRORA) without neovascularization are synonymous with GA. In the trial, patient eyes receiving pegcetacoplan possessed lower rates of progression from iRORA to cRORA. As such, although findings did not achieve statistical significance, pegcetacoplan may prevent the incidence of GA [66]. However, in the FILLY study, new-onset wet AMD was more prevalent in the pegcetacoplan groups compared with the sham arm. Thus, participants who developed wet AMD with pegcetacoplan discontinued treatment [65].

An additional post hoc analysis of the FILLY trial determined 20.9% (18 of 86), 8.9% (7 of 79), and 1.2% (1 of 81) of study eyes developed wet AMD in the pegcetacoplan monthly, pegcetacoplan EOM, and sham groups, respectively, throughout the 18-month study course. Importantly, eyes that eventually developed wet AMD demonstrated increased baseline prevalence of both the double-layer sign (DLS), a finding suggestive of type I macular neovascularization (MNV), and wet AMD in the fellow eye [67]. Irrespective of the inclusion or exclusion of data for patients with new onset wet AMD, the study's primary endpoint of reducing GA growth rate was still attained in the FILLY trial. Furthermore, an independent safety monitoring committee permitted completion of the study, as the exudations did not appreciably impact visual acuity [67].

Due to its acceptable safety profile and statistically significant reductions in GA lesion expansion, pegcetacoplan was cleared for multicenter, randomized phase III trials, labeled DERBY (NCT03525600) and OAKS (NCT03525613), both of which have completed enrollment (n = 621 and n = 638, respectively) [68], but remain ongoing [65,67]. During these investigations, patients who develop wet AMD in the study eyes will be concomitantly treated with anti-VEGF therapy, enabling subjects to continue receiving pegcetacoplan. This protocol will facilitate further delineation of the advantages and disadvantages of continued pegcetacoplan administration in addition to potential linkages between pegcetacoplan and new onset wet AMD [65,67]. Preliminary data at 18 months from the DERBY and OAKS trial have been encouraging. A combined analysis of the trials showed a 13% GA growth rate reduction in foveal lesions in the monthly ($p = 0.0070$) and EOM ($p = 0.0069$) pegcetacoplan groups relative to sham. Moreover, the GA growth rate of extra-foveal lesions was reduced by 21% ($p = 0.0006$) and 26% ($p < 0.0001$) in the EOM and monthly groups, respectively. Regarding new onset wet AMD in the combined studies, rates of 9.5% (40 of 419), 6.2% (26 of 420), and 2.9% (12 of 417) were noted at month 18 in the monthly, EOM, and sham arms, respectively. Across the pegcetacoplan monthly and EOM arms, 4 cases of endophthalmitis have been reported, translating to an infection rate of 0.044% per injection [69].

Another planned investigation is the GALE phase III trial (NCT04770545), a multicenter, non-randomized extension study that will further evaluate the safety and efficacy of pegcetacoplan using patients from the initial phase I trial (NCT03777332) and patients who complete 24 months of treatment from the DERBY and OAKS trials. Recruitment of 1200 patients is anticipated.

6.2.3. POT-4 (AL-78898A)

POT-4 (AL-78898A) is a compstatin analog originally developed by Potentia Pharmaceuticals that functions as a C3 inhibitor and was the first complement inhibitor tested in humans with AMD [60,70]. Phase I trial (NCT00473928) results revealed no adverse drug events (ADEs) or serious adverse events (SAEs) with doses of up to 450 mg for patients with wet AMD [70]. Based on the tolerability of POT-4, a multicenter, randomized phase II trial (NCT01603043) with 10 patients was conducted to assess its efficacy in reducing GA lesion growth among individuals with dry AMD. However, the study was terminated prematurely, as four of seven participants (57.14%) in the POT-4 group developed drug product deposits in the eye.

6.2.4. NGM621

NGM621 (NGM Biopharmaceuticals, San Francisco, CA, USA) is a humanized immunoglobulin G1 monoclonal antibody formulated as a C3 inhibitor. A phase I trial (NCT04014777) included 15 patients with GA secondary to dry AMD [71]. Participants were treated with either single-ascending doses (2 mg, 7.5 mg, 15 mg) of NGM621 or two doses of 15 mg delivered 4 weeks apart. Monitoring occurred for 12 weeks within all cohorts, and results indicated an appropriate safety profile, with no SAEs, ADEs, or new-onset CNV reported.

These promising findings resulted in the development of a multicenter, randomized phase II trial entitled CATALINA (NCT04465955) investigating the safety and efficacy of 15 mg of NGM621 administered every 4 or 8 weeks compared to sham throughout a 52-week interval [71]. The primary outcome is the rate of change in the GA lesion area over the trial period. In total, 320 participants with GA are currently participating in this ongoing trial.

6.3. C5

6.3.1. Eculizumab

Eculizumab (Alexion Pharmaceuticals, Inc) is a humanized monoclonal antibody directed against C5 [72,73]. Systemic eculizumab is currently approved for the treatment of paroxysmal nocturnal hemoglobinuria and atypical hemolytic uremic syndrome. Thus, considering the evidence for the relationship of the complement cascade with AMD, the COMPLETE study (NCT00935883) was designed [72,73].

The single-center, randomized phase II trial evaluated the safety and efficacy of intravenous (IV) eculizumab for the treatment of GA among individuals with dry AMD. Overall, 60 patients were enrolled, with patients stratified into two distinct cohorts characterized by the presence of GA or the presence of drusen but the absence of GA (indicative of intermediate dry AMD). Each cohort of 30 patients was subsequently enrolled in a 2:1 ratio to IV eculizumab or placebo. The first 10 patients in the eculizumab group were provided with a low-dose regimen of 600 mg weekly for 4 weeks, followed by 900 mg every 2 weeks until week 24 of the study. The remaining 10 patients were provided with the high-dose regimen of 900 mg weekly for 4 weeks, followed by 1200 mg every 2 weeks until week 24. In the GA cohort, therapeutic efficacy was evaluated through a change in the square root of the lesion area at week 26. Although eculizumab demonstrated acceptable tolerability, average differences in the square root of GA area at 26 weeks were 0.19 ($\pm$0.12) mm and 0.18 ($\pm$0.15) mm in the eculizumab and placebo groups, respectively ($p = 0.96$) [72]. In the drusen cohort, therapeutic efficacy was evaluated through the presence of a 50% reduction in drusen volume at week 26. No patients treated with eculizumab achieved this primary outcome [73].

6.3.2. Avacincaptad Pegol (Zimura® [ARC1905])

Avacincaptad pegol (Zimura®), by Iveric Bio (Cranbury, NJ, USA), is a PEGylated RNA aptamer that operates as a C5 cleavage inhibitor, thereby impeding the complement cascade irrespective of the initial activation pathway. This agent was evaluated in the phase II/III GATHER1 trial (NCT02686658), a multicenter, randomized investigation. Randomization was conducted in two stages. In total, 77 part I participants were randomized in a 1:1:1 ratio to injections of 1 mg of avacincaptad pegol, 2 mg of avacincaptad pegol, and sham. Then, 209 part II participants were randomized in a 1:2:2 ratio to 2 mg of avacincaptad pegol, 4 mg of avacincaptad pegol, and sham. Relative to sham, subjects receiving 2 mg of avacincaptad pegol and 4 mg of avacincaptad pegol experienced a 27.4% ($p = 0.0072$) and 27.8% ($p = 0.0051$) reduction, respectively, in the mean rate of GA growth [74].

Avacincaptad pegol was generally well-tolerated, as no ADEs or SAEs were noted following 12 months of treatment. Compared to pegcetacoplan, a decreased incidence of endophthalmitis and CNV was reported with avacincaptad pegol [74]. Ongoing phase III trials will be beneficial to further exploring this phenomenon [65,74]. The 18-month results from the GATHER1 clinical trial continue to validate the safety and efficacy of avacincaptad pegol [75]. Compared to sham, the cohort treated with 2 mg of avacincaptad pegol demonstrated a 28% reduction ($p < 0.0014$) in GA growth and the cohort treated with 4 mg of avacincaptad pegol demonstrated a 30% reduction ($p < 0.0021$) from baseline to 18-month follow-up. No ADEs were reported up to 18 months post-administration with avacincaptad pegol [75].

Based on the efficacy and safety of avacincaptad pegol, another phase III trial, the multicenter, randomized GATHER2 study (NCT04435366), which compares 2 mg of avacincaptad pegol with sham, is currently being conducted. If data from the 448 patients currently enrolled in GATHER2 similarly reveal statistically significant reductions in GA growth rate with avacincaptad pegol, the treatment will have the two phase III trials necessary for FDA approval application [74].

6.3.3. Tesidolumab (LFG316)

Tesidolumab (LFG316), a monoclonal C5 inhibitor developed by Novartis Pharmaceuticals (Basel, Switzerland), was investigated in a multicenter, randomized phase II clinical trial (NCT01527500). In the study, 158 patients were enrolled and separated into two groups. Part A examined the safety and efficacy of multiple 5 mg IVT injections of LFG316 relative to sham every 28 days over the course of 505 days. Part B examined the safety and pharmacokinetic properties of a single 10 mg IVT injection. Results revealed a relatively benign safety profile, with no demonstrable improvement in the primary outcome of GA lesion growth or the secondary outcome of BCVA.

6.4. *Complement Factor B*

IONIS-FB-lrx

In more recent years, therapies have emerged to alter gene expression rather than target proteins. IONIS-FB-lrx (Ionis Pharmaceuticals, Carlsbad, CA, USA) is a novel antisense oligonucleotide (ASO) targeting the gene encoding complement factor B (CFB), a moiety of the alternative complement pathway. When delivered subcutaneously, IONIS-FB-lrx reduced circulating levels of CFB in a dose-dependent manner among 54 healthy participants of a phase I trial [76]. No notable adverse effects were reported. Given these results, the multicenter, randomized phase II GOLDEN trial (NCT03815825) has been initiated to evaluate the influence of IONIS-FB-lrx administration on the progression of GA lesion size [76]. Anticipated recruitment is 330 patients.

6.5. *Complement Factor D*

Lampalizumab (FCFD4514S)

Complement factor D (CFD), a rate-limiting enzyme of the alternative pathway, converts proconvertases into active C3 and C5 convertases. Consequently, inhibition of CFD has surfaced as an attractive therapeutic option for dry AMD [77]. Lampalizumab, designed by Genentech (San Francisco, CA, USA), is a humanized monoclonal antibody with an antigen-binding fragment that inhibits complement factor D.

In a phase I trial (NCT00973011), 18 participants received 10 mg of IVT lampalizumab. The agent had an acceptable safety profile without notable ocular or systemic ADEs or SAEs. To expand these findings, the MAHALO phase II trial (NCT01229215) was conducted to evaluate lampalizumab's efficacy in reducing GA progression [78]. Monthly treatment with lampalizumab resulted in a statistically significant 20% reduction ($p = 0.117$) in mean GA lesion enlargement relative to sham. Patients with high-risk complement factor I (CFI) alleles displayed a greater response to monthly lampalizumab with a 44% reduction ($p = 0.0037$) in GA atrophy progression compared to sham. Similar to phase I findings, lampalizumab was well-tolerated [78].

These findings encouraged the design of the CHROMA (NCT02247479) and SPECTRI (NCT02247531) multicenter, randomized phase III trials to further delineate lampalizumab's effects on GA enlargement [79]. Both investigations randomized 1881 total participants in a 2:1:2:1 configuration to 10 mg of IVT lampalizumab every 4 weeks, sham every 4 weeks, 10 mg lampalizumab every 6 weeks, or sham every 6 weeks, over a period of 96 weeks. Efficacy was assessed as a change in GA area from baseline to week 48 of the study. Despite acceptable tolerability, lampalizumab was determined not to be efficacious for the treatment of GA. 48 weeks post-administration, the difference in means (lampalizumab vs. Sham pooled) was 0.071 mm^2 ($p = 0.25$) for lampalizumab every 4 weeks and 0.070 mm^2 ($p = 0.25$)

for lampalizumab every 6 weeks. In contrast to the MAHALO study, both CHROMA and SPECTRI identified no relationship between CFI risk alleles and GA progression. Random chance and small sample size in the MAHALO study were posited to have contributed to this discrepancy [79].

Visual function decline among patients in the CHROMA and SPECTRI trials was examined utilizing several metrics, including best corrected visual acuity (BCVA), low-luminance visual acuity (LLVA), maximum reading speed, microperimetry, and patient-reported outcomes (PRO), as quantified via the Functional Reading Independence (FRI) Index and the National Eye Institute Visual Function Questionnaire (NEI VFQ-25). Results of the analysis revealed no statistically significant difference in visual function assessments across all treatment groups at week 48 of the phase III trials [80].

6.6. Complement Factor H

6.6.1. AdCAGfH

CFH downregulates the alternative complement cascade by preventing C3 convertase formation from c3b; therefore, it has been postulated as a potential therapeutic for dry AMD. The injection of AdCAGfH, an adeno-associated virus (AAV) containing murine factor H, demonstrated an attenuation of C3-induced retinal inflammation [81]. This gene therapy is currently under consideration for human subjects [82].

6.6.2. GEM103

GEM103 is a recombinant human CFH developed by Gemini Therapeutics (Cambridge, MA, USA). In a phase I clinical trial (NCT04246866) comprised of 12 participants, GEM103 demonstrated minimal adverse effects and a dose-dependent increase in levels of CFH [83]. Significantly, no CNV was observed, a phenomenon previously reported in the literature in association with repeated administration of IVT complement inhibitors [67].

Interim findings from the subsequent multicenter, randomized ReGAtta phase IIa study (NCT04643886) have suggested positive alterations in biomarkers linked to the complement-mediated pathogenesis of dry AMD. The safety profile of GEM103 remained relatively benign, with adverse events reported among 16 of 62 volunteers, of which one case of mild iritis was directly related to the compound [84]. Since the ReGAtta trial achieved its primary endpoint of evaluating the safety and tolerability of GEM103, the trial was terminated [85].

6.7. Complement Factor I

GT005

CFI is a regulator of the alternative pathway. With the assistance of its cofactors, it cleaves C3b into pro-inflammatory ic3b, which is subsequently cleaved into inert c3dg. Thus, the protein is crucial for modulating the activation of the alternative pathway. Importantly, experimental investigations confirmed the ability of AAV-containing CFI cDNA to induce CFI protein expression in human RPE lines and murine retina, thereby stimulating the development of GT005 by Gyroscope Therapeutics.

The multicenter, randomized FOCUS phase I/II trial (NCT03846193), which is currently recruiting patients (65 anticipated), is evaluating this agent among individuals with GA secondary to dry AMD [86]. Interim data for 31 patients demonstrated a significant increase in vitreous CFI and a significant reduction in C3 and its cleavage products following administration of GT005. No SAEs were reported in the treatment cohort [87].

Additional multicenter, randomized phase II trials, entitled HORIZON (NCT04566445) and EXPLORE (NCT04437368) are evaluating a two-dose single injection of GT005 in patents with GA from dry AMD. The HORIZON and EXPLORE trials will include patients previously genotyped for rare CFI genetic variants, enabling elucidation of the interaction between the therapeutic's effects and the presence of CFI mutations on the advancement of GA. Anticipated recruitment is currently 250 and 75 patients, respectively.

6.8. Membrane Attack Complex (MAC)

AAVCAGsCD59 (HMR59)

Regulators of the complement cascade are emerging targets for investigational compounds. AAVCAGsCD59, also denoted HMR59, is an intravitreally injected AAV vector developed by Hemera Biosciences that induces the formation of CD59, thereby preventing the binding of C9 required for the formation of a complete MAC [88]. Pre-clinically, HMR59 demonstrated efficacy in attenuating laser-induced CNV among murine models [89].

The interim results from a non-randomized phase I trial involving 17 eyes with dry AMD (NCT03144999) have indicated promise, with acceptable tolerability and no dose-limiting toxicity. While adverse events were limited, it is important to note that four eyes developed mild inflammation that subsequently resolved with topical steroid therapy or observation. Two of these four patients additionally required topical medication to reduce intraocular pressure. Encouragingly, although not powered to illustrate efficacy, findings from the investigation revealed that 9 of 11 patients had decreased GA progression relative to historical controls [90]. As such, a multicenter, randomized phase II trial is currently planned (NCT04358471) to evaluate the therapeutic's effect on the expansion of the GA area.

6.9. Properdin

CLG561

Properdin is a positive regulator of the alternative complement pathway, preventing the degradation of C3 and C5 convertases. Thus, CLG561, an anti-properdin antibody, was developed as a potential therapeutic by Alcon Research [91]. Results from a phase I trial (NCT01835015) showed IVT doses of CLG561 up to 10 mg were safe and well-tolerated, with no systemic or ocular ADEs reported [92].

Subsequently, a multicenter, randomized phase II trial (NCT02515942) was conducted to evaluate CLG561, both individually and in combination with LFG316 (tesidolumab), against sham in 114 participants with GA. Twelve total injections were administered per group, with injections occurring every 28 days. Change in GA lesion size from baseline to day 337 was the primary outcome. Neither CLG561 monotherapy ($p = 0.1019$) nor CLG561 plus LFG316 ($p = 0.5987$) significantly reduced GA lesion growth relative to sham at day 337.

7. Challenges and Future Directions

Translation of experimental research for dry AMD is partially impeded by experimental models that are incapable of elucidating and replicating the complex pathogenic processes in human eyes [36,93]. As the disease is a multifactorial process, it is essential for these models to incorporate variables such as oxidative stress, free radicals, and peroxidation to ensure that results are applicable to a clinical environment, a process that has hitherto remained elusive.

In recent years, several experimental animal models have been established to identify characteristics of dry AMD similar to the pathology observed in humans. Traditionally, rodents have been preferred to primates because of their short life cycle, rapid disease progression, and genetic similarity to humans. One challenge associated with the rodent model is the crucial differences in the anatomical structures of the eye. While dry AMD primarily affects central vision through degeneration of the fovea and macula, these structures are absent in rodents [36,94,95]. Apart from anatomy, other key AMD pathology such as drusen, RPE atrophy, and CNV are not exhibited in rodent models [95,96]. These differences preclude a comprehensive understanding of the mechanism of AMD, thereby hindering the discovery of novel therapeutics. Advancements in this arena are currently being propelled by the development of three-dimensional in vitro models using human cells to represent the pathology entwined within BM and the RPE layer [97]. Such an innovation would provide essential insights into the biomechanical performance of the retina and replace current in vivo animal models that provide an incomplete understanding of the disease.

Nevertheless, even where experimental research may be translated into the development of clinical compounds, modulating the complement cascade creates challenges for a myriad of reasons. Interference in this immunogenic process may result in innumerable adverse effects, an essential consideration for therapeutics administered over an extended period for the management of chronic diseases. To ameliorate this concern, pharmacologic inhibition should exclusively target the dysregulated components of the complement cascade, thus limiting the associated systemic effects. However, extensive exploration remains to delineate the mechanisms of disease progression in order to successfully produce such compounds.

Furthermore, there is notable intrapopulation variation in the expression of genetic risk factors contributory to the pathogenesis of dry AMD [15,20,46,98,99]. Because complement dysregulation comprises merely a single component of this multifactorial condition, the selection of individuals possessing a particular predisposition to complement-mediated disease progression would enable the development of personalized treatments for patients who would extract the most benefit. Identifying specific genetic variants associated with increased risk (e.g., rs1061170 variant of CFH) or calculating aggregate genetic risk profiles (from multiple variants) to stratify patients are potential approaches to facilitating more efficacious clinical trials. Alternatively, investigational compounds targeting pathways implicated in extracellular matrix remodeling, lipid metabolism, or oxidative stress would be of greater suitability for individuals not possessing these critical genetic risk factors [100–103].

Clinically, patient adherence to a structured treatment regimen is a substantial challenge that persists across the ophthalmic landscape [104,105]. To circumvent this barrier, innovative modalities of ocular delivery such as biodegradable polymeric implants, colloids, and hydrogels are currently being explored as potential solutions [106–109]. These sophisticated alternatives have the capacity to rapidly accelerate the adoption of dry AMD therapeutics.

8. Conclusions

Despite the absence of available therapeutics for patients with GA secondary to dry AMD, there remains significant hope that one of the numerous clinical trials will produce an effective treatment option. The probable association of the complement cascade with the pathogenesis of dry AMD, as evidenced by genetic, experimental, and clinical investigations, emphasizes the integral participation of inflammation in the development and progression of the disease. This pathway offers a prolific target for the synthesis of novel therapeutics. As a continued exploration of the linkages between the complement cascade and GA secondary to dry AMD occurs, our understanding of the disease will expand beyond its current state, enabling further identification of treatments for patients.

Author Contributions: Conceptualization, P.N.P., P.A.P. and V.S.; methodology, P.N.P. and P.A.P.; formal analysis, P.N.P. and P.A.P.; data curation, P.N.P., P.A.P., M.R.L., I.B.-T. and H.A.; writing—original draft preparation, P.N.P., P.A.P., M.R.L., I.B.-T. and H.A.; writing—review and editing, V.S.; visualization, V.S.; supervision, V.S.; funding acquisition, V.S. All authors have read and agreed to the published version of the manuscript.

Funding: This research received no external funding.

Institutional Review Board Statement: Not applicable.

Informed Consent Statement: Not applicable.

Data Availability Statement: No new data were created or analyzed in this study. Data sharing is not applicable to this article.

Conflicts of Interest: The authors declare no conflict of interest.

References

1. Bourne, R.R.; Jonas, J.B.; Bron, A.M.; Cicinelli, M.V.; Das, A.; Flaxman, S.R.; Friedman, D.S.; Keeffe, J.E.; Kempen, J.H.; Leasher, J. Prevalence and causes of vision loss in high-income countries and in Eastern and Central Europe in 2015: Magnitude, temporal trends and projections. *Br. J. Ophthalmol.* **2018**, *102*, 575–585. [CrossRef] [PubMed]
2. Wong, W.L.; Su, X.; Li, X.; Cheung, C.M.G.; Klein, R.; Cheng, C.-Y.; Wong, T.Y. Global prevalence of age-related macular degeneration and disease burden projection for 2020 and 2040: A systematic review and meta-analysis. *Lancet Glob. Health* **2014**, *2*, e106–e116. [CrossRef]
3. Kim, S.; Park, S.J.; Byun, S.J.; Park, K.H.; Suh, H.S. Incremental economic burden associated with exudative age-related macular degeneration: A population-based study. *BMC Health Serv. Res.* **2019**, *19*, 828. [CrossRef]
4. Sadda, S.R.; Guymer, R.; Holz, F.G.; Schmitz-Valckenberg, S.; Curcio, C.A.; Bird, A.C.; Blodi, B.A.; Bottoni, F.; Chakravarthy, U.; Chew, E.Y.; et al. Consensus Definition for Atrophy Associated with Age-Related Macular Degeneration on OCT: Classification of Atrophy Report 3. *Ophthalmology* **2018**, *125*, 537–548. [CrossRef] [PubMed]
5. Bird, A.C.; Bressler, N.M.; Bressler, S.B.; Chisholm, I.H.; Coscas, G.; Davis, M.D.; de Jong, P.T.; Klaver, C.C.; Klein, B.E.; Klein, R.; et al. An international classification and grading system for age-related maculopathy and age-related macular degeneration. The International ARM Epidemiological Study Group. *Surv. Ophthalmol.* **1995**, *39*, 367–374. [CrossRef]
6. Ferris, F.L.; Davis, M.D.; Clemons, T.E.; Lee, L.Y.; Chew, E.Y.; Lindblad, A.S.; Milton, R.C.; Bressler, S.B.; Klein, R. A simplified severity scale for age-related macular degeneration: AREDS Report No. 18. *Arch. Ophthalmol.* **2005**, *123*, 1570–1574. [CrossRef]
7. Ferris, F.L., 3rd; Wilkinson, C.P.; Bird, A.; Chakravarthy, U.; Chew, E.; Csaky, K.; Sadda, S.R. Clinical classification of age-related macular degeneration. *Ophthalmology* **2013**, *120*, 844–851. [CrossRef]
8. Keane, P.A.; Liakopoulos, S.; Ongchin, S.C.; Heussen, F.M.; Msutta, S.; Chang, K.T.; Walsh, A.C.; Sadda, S.R. Quantitative subanalysis of optical coherence tomography after treatment with ranibizumab for neovascular age-related macular degeneration. *Investig. Ophthalmol. Vis. Sci.* **2008**, *49*, 3115–3120. [CrossRef]
9. Horner, F.; Lip, P.L.; Mohammed, B.R.; Fusi-Rubiano, W.; Gokhale, E.; Mushtaq, B.; Chavan, R. Comparing Effectiveness of Three Different Anti-VEGF Treatment Regimens for Neovascular Age-Related Macular Degeneration: Two Years' Real-World Clinical Outcomes. *Clin. Ophthalmol.* **2021**, *15*, 1703–1713. [CrossRef]
10. Arrigo, A.; Saladino, A.; Aragona, E.; Mercuri, S.; Introini, U.; Bandello, F.; Parodi, M.B. Different Outcomes of Anti-VEGF Treatment for Neovascular AMD according to Neovascular Sutypes and Baseline Features: 2-Year Real-Life Clinical Outcomes. *BioMed Res. Int.* **2021**, *2021*, 5516981. [CrossRef]
11. Brown, G.C.; Brown, M.M.; Rapuano, S.; Boyer, D. Cost-Utility Analysis of VEGF Inhibitors for Treating Neovascular Age-Related Macular Degeneration. *Am. J. Ophthalmol.* **2020**, *218*, 225–241. [CrossRef]
12. Stanton, C.M.; Yates, J.R.; den Hollander, A.I.; Seddon, J.M.; Swaroop, A.; Stambolian, D.; Fauser, S.; Hoyng, C.; Yu, Y.; Atsuhiro, K.; et al. Complement factor D in age-related macular degeneration. *Investig. Ophthalmol. Vis. Sci.* **2011**, *52*, 8828–8834. [CrossRef]
13. van de Ven, J.P.H.; Nilsson, S.C.; Tan, P.L.; Buitendijk, G.H.S.; Ristau, T.; Mohlin, F.C.; Nabuurs, S.B.; Schoenmaker-Koller, F.E.; Smailhodzic, D.; Campochiaro, P.A.; et al. A functional variant in the CFI gene confers a high risk of age-related macular degeneration. *Nat. Genet.* **2013**, *45*, 813–817. [CrossRef]
14. Maugeri, A.; Barchitta, M.; Agodi, A. The association between complement factor H rs1061170 polymorphism and age-related macular degeneration: A comprehensive meta-analysis stratified by stage of disease and ethnicity. *Acta Ophthalmol.* **2019**, *97*, e8–e21. [CrossRef]
15. Heesterbeek, T.J.; Lorés-Motta, L.; Hoyng, C.B.; Lechanteur, Y.T.; den Hollander, A.I. Risk factors for progression of age-related macular degeneration. *Ophthalmic Physiol. Opt.* **2020**, *40*, 140–170. [CrossRef]
16. Colijn, J.M.; Buitendijk, G.H.S.; Prokofyeva, E.; Alves, D.; Cachulo, M.L.; Khawaja, A.P.; Cougnard-Gregoire, A.; Merle, B.M.J.; Korb, C.; Erke, M.G.; et al. Prevalence of Age-Related Macular Degeneration in Europe: The Past and the Future. *Ophthalmology* **2017**, *124*, 1753–1763. [CrossRef]
17. Joachim, N.; Mitchell, P.; Kifley, A.; Rochtchina, E.; Hong, T.; Wang, J.J. Incidence and progression of geographic atrophy: Observations from a population-based cohort. *Ophthalmology* **2013**, *120*, 2042–2050. [CrossRef]
18. Kuan, V.; Warwick, A.; Hingorani, A.; Tufail, A.; Cipriani, V.; Burgess, S.; Sofat, R.; Consortium, I.A.G. Association of Smoking, Alcohol Consumption, Blood Pressure, Body Mass Index, and Glycemic Risk Factors With Age-Related Macular Degeneration: A Mendelian Randomization Study. *JAMA Ophthalmol.* **2021**, *139*, 1299–1306. [CrossRef]
19. Seddon, J.M.; Widjajahakim, R.; Rosner, B. Rare and Common Genetic Variants, Smoking, and Body Mass Index: Progression and Earlier Age of Developing Advanced Age-Related Macular Degeneration. *Investig. Ophthalmol. Vis. Sci.* **2020**, *61*, 32. [CrossRef]
20. Yu, Y.; Reynolds, R.; Rosner, B.; Daly, M.J.; Seddon, J.M. Prospective assessment of genetic effects on progression to different stages of age-related macular degeneration using multistate Markov models. *Investig. Ophthalmol. Vis. Sci.* **2012**, *53*, 1548–1556. [CrossRef]
21. Agrón, E.; Mares, J.; Chew, E.Y.; Keenan, T.D.L. Adherence to a Mediterranean Diet and Geographic Atrophy Enlargement Rate: Age-Related Eye Disease Study 2 Report 29. *Ophthalmol. Retin.* **2022**. [CrossRef]
22. Chapman, N.A.; Jacobs, R.J.; Braakhuis, A.J. Role of diet and food intake in age-related macular degeneration: A systematic review. *Clin. Exp. Ophthalmol.* **2019**, *47*, 106–127. [CrossRef] [PubMed]

23. Yip, J.L.; Khawaja, A.P.; Chan, M.P.; Broadway, D.C.; Peto, T.; Tufail, A.; Luben, R.; Hayat, S.; Bhaniani, A.; Wareham, N.J.; et al. Cross Sectional and Longitudinal Associations between Cardiovascular Risk Factors and Age Related Macular Degeneration in the EPIC-Norfolk Eye Study. *PLoS ONE* **2015**, *10*, e0132565. [CrossRef] [PubMed]
24. Zhou, H.; Zhang, H.; Yu, A.; Xie, J. Association between sunlight exposure and risk of age-related macular degeneration: A meta-analysis. *BMC Ophthalmol.* **2018**, *18*, 331. [CrossRef] [PubMed]
25. Sterling, J.K.; Baumann, B.; Foshe, S.; Voigt, A.; Guttha, S.; Alnemri, A.; McCright, S.J.; Li, M.; Zauhar, R.J.; Montezuma, S.R.; et al. Inflammatory adipose activates a nutritional immunity pathway leading to retinal dysfunction. *Cell Rep.* **2022**, *39*, 110942. [CrossRef] [PubMed]
26. Kalu, K.; McMonnies, C.; Arcot, J. Dietary and Non-Dietary Determinants Associated with Changes in Red Blood Cell Omega-3-Essential Fatty Acids and Serum Lutein Status in a Selected Adult Population. *Curr. Dev. Nutr.* **2022**, *6*, 295. [CrossRef]
27. Fritsche, L.G.; Igl, W.; Bailey, J.N.; Grassmann, F.; Sengupta, S.; Bragg-Gresham, J.L.; Burdon, K.P.; Hebbring, S.J.; Wen, C.; Gorski, M.; et al. A large genome-wide association study of age-related macular degeneration highlights contributions of rare and common variants. *Nat. Genet.* **2016**, *48*, 134–143. [CrossRef]
28. Yan, Q.; Ding, Y.; Liu, Y.; Sun, T.; Fritsche, L.G.; Clemons, T.; Ratnapriya, R.; Klein, M.L.; Cook, R.J.; Liu, Y.; et al. Genome-wide analysis of disease progression in age-related macular degeneration. *Hum. Mol. Genet.* **2018**, *27*, 929–940. [CrossRef]
29. Tang, M.; Li, A.; Wu, M.; Chen, X.; Xiong, X.; Zhou, Z.; Liu, D. rs10490924 surrounding HTRA1/ARMS2 regulates the susceptibility of age-related macular degeneration. *J. Recept. Signal Transduct.* **2021**, *41*, 188–195. [CrossRef]
30. Micklisch, S.; Lin, Y.; Jacob, S.; Karlstetter, M.; Dannhausen, K.; Dasari, P.; von der Heide, M.; Dahse, H.-M.; Schmölz, L.; Grassmann, F.; et al. Age-related macular degeneration associated polymorphism rs10490924 in ARMS2 results in deficiency of a complement activator. *J. Neuroinflamm.* **2017**, *14*, 4. [CrossRef]
31. Ratnapriya, R.; Acar, İ.E.; Geerlings, M.J.; Branham, K.; Kwong, A.; Saksens, N.T.M.; Pauper, M.; Corominas, J.; Kwicklis, M.; Zipprer, D.; et al. Family-based exome sequencing identifies rare coding variants in age-related macular degeneration. *Hum. Mol. Genet.* **2020**, *29*, 2022–2034. [CrossRef]
32. Hu, M.L.; Quinn, J.; Xue, K. Interactions between Apolipoprotein E Metabolism and Retinal Inflammation in Age-Related Macular Degeneration. *Life* **2021**, *11*, 635. [CrossRef]
33. Liutkeviciene, R.; Vilkeviciute, A.; Kriauciuniene, L.; Banevicius, M.; Budiene, B.; Stanislovaitiene, D.; Zemaitiene, R.; Deltuva, V.P. Association of genetic variants at CETP, AGER, and CYP4F2 locus with the risk of atrophic age-related macular degeneration. *Mol. Genet. Genom. Med.* **2020**, *8*, e1357. [CrossRef]
34. Bhattarai, N.; Hytti, M.; Reinisalo, M.; Kaarniranta, K.; Mysore, Y.; Kauppinen, A. Hydroquinone predisposes for retinal pigment epithelial (RPE) cell degeneration in inflammatory conditions. *Immunol. Res.* **2022**. [CrossRef]
35. Chang, Y.-Y.; Lee, Y.-J.; Hsu, M.-Y.; Wang, M.; Tsou, S.-C.; Chen, C.-C.; Lin, J.-A.; Hsiao, Y.-P.; Lin, H.-W. Protective Effect of Quercetin on Sodium Iodate-Induced Retinal Apoptosis through the Reactive Oxygen Species-Mediated Mitochondrion-Dependent Pathway. *Int. J. Mol. Sci.* **2021**, *22*, 4056. [CrossRef]
36. Soundara Pandi, S.P.; Ratnayaka, J.A.; Lotery, A.J.; Teeling, J.L. Progress in developing rodent models of age-related macular degeneration (AMD). *Exp. Eye Res.* **2021**, *203*, 108404. [CrossRef]
37. Nagata, K.; Hishikawa, D.; Sagara, H.; Saito, M.; Watanabe, S.; Shimizu, T.; Shindou, H. Lysophosphatidylcholine acyltransferase 1 controls mitochondrial reactive oxygen species generation and survival of retinal photoreceptor cells. *J. Biol. Chem.* **2022**, *298*, 101958. [CrossRef]
38. Hsu, M.-Y.; Hsiao, Y.-P.; Lin, Y.-T.; Chen, C.; Lee, C.-M.; Liao, W.-C.; Tsou, S.-C.; Lin, H.-W.; Chang, Y.-Y. Quercetin Alleviates the Accumulation of Superoxide in Sodium Iodate-Induced Retinal Autophagy by Regulating Mitochondrial Reactive Oxygen Species Homeostasis through Enhanced Deacetyl-SOD2 via the Nrf2-PGC-1α-Sirt1 Pathway. *Antioxidants* **2021**, *10*, 1125. [CrossRef]
39. Eells, J.T. Mitochondrial Dysfunction in the Aging Retina. *Biology* **2019**, *8*, 31. [CrossRef]
40. Lazzarini, R.; Nicolai, M.; Lucarini, G.; Pirani, V.; Mariotti, C.; Bracci, M.; Mattioli-Belmonte, M. Oxidative stress in retinal pigment epithelium impairs stem cells: A vicious cycle in age-related macular degeneration. *Mol. Cell. Biochem.* **2022**, *477*, 67–77. [CrossRef]
41. Anderson, D.H.; Mullins, R.F.; Hageman, G.S.; Johnson, L.V. A role for local inflammation in the formation of drusen in the aging eye. *Am. J. Ophthalmol.* **2002**, *134*, 411–431. [CrossRef]
42. Wang, J.; Zhang, H.; Ji, J.; Wang, L.; Lv, W.; He, Y.; Li, X.; Feng, G.; Chen, K. A histological study of atherosclerotic characteristics in age-related macular degeneration. *Heliyon* **2022**, *8*, e08973. [CrossRef] [PubMed]
43. van Lookeren Campagne, M.; LeCouter, J.; Yaspan, B.L.; Ye, W. Mechanisms of age-related macular degeneration and therapeutic opportunities. *J. Pathol.* **2014**, *232*, 151–164. [CrossRef] [PubMed]
44. Handa, J.T.; Bowes Rickman, C.; Dick, A.D.; Gorin, M.B.; Miller, J.W.; Toth, C.A.; Ueffing, M.; Zarbin, M.; Farrer, L.A. A systems biology approach towards understanding and treating non-neovascular age-related macular degeneration. *Nat. Commun.* **2019**, *10*, 3347. [CrossRef]
45. Kim, S.-Y.; Kambhampati, S.P.; Bhutto, I.A.; McLeod, D.S.; Lutty, G.A.; Kannan, R.M. Evolution of oxidative stress, inflammation and neovascularization in the choroid and retina in a subretinal lipid induced age-related macular degeneration model. *Exp. Eye Res.* **2021**, *203*, 108391. [CrossRef]
46. Garcia-Garcia, J.; Usategui-Martin, R.; Sanabria, M.R.; Fernandez-Perez, E.; Telleria, J.J.; Coco-Martin, R.M. Pathophysiology of Age-Related Macular Degeneration. Implications for Treatment. *Ophthalmic Res.* **2022**. [CrossRef]

47. Fleckenstein, M.; Mitchell, P.; Freund, K.B.; Sadda, S.; Holz, F.G.; Brittain, C.; Henry, E.C.; Ferrara, D. The Progression of Geographic Atrophy Secondary to Age-Related Macular Degeneration. *Ophthalmology* **2018**, *125*, 369–390. [CrossRef]
48. Schmitz-Valckenberg, S.; Sadda, S.; Staurenghi, G.; Chew, E.Y.; Fleckenstein, M.; Holz, F.G. Geographic Atrophy: Semantic Considerations and Literature Review. *Retina* **2016**, *36*, 2250–2264. [CrossRef]
49. Lorthiois, E.; Anderson, K.; Vulpetti, A.; Rogel, O.; Cumin, F.; Ostermann, N.; Steinbacher, S.; Mac Sweeney, A.; Delgado, O.; Liao, S.-M. Discovery of highly potent and selective small-molecule reversible factor D inhibitors demonstrating alternative complement pathway inhibition in vivo. *J. Med. Chem.* **2017**, *60*, 5717–5735. [CrossRef]
50. Desai, D.; Dugel, P.U. Complement cascade inhibition in geographic atrophy: A review. *Eye* **2022**, *36*, 294–302. [CrossRef]
51. Bora, P.S.; Sohn, J.H.; Cruz, J.M.; Jha, P.; Nishihori, H.; Wang, Y.; Kaliappan, S.; Kaplan, H.J.; Bora, N.S. Role of complement and complement membrane attack complex in laser-induced choroidal neovascularization. *J. Immunol.* **2005**, *174*, 491–497. [CrossRef]
52. Natoli, R.; Fernando, N.; Jiao, H.; Racic, T.; Madigan, M.; Barnett, N.L.; Chu-Tan, J.A.; Valter, K.; Provis, J.; Rutar, M. Retinal Macrophages Synthesize C3 and Activate Complement in AMD and in Models of Focal Retinal Degeneration. *Investig. Ophthalmol. Vis. Sci.* **2017**, *58*, 2977–2990. [CrossRef]
53. Wu, J.; Sun, X. Complement system and age-related macular degeneration: Drugs and challenges. *Drug Des. Dev. Ther.* **2019**, *13*, 2413–2425. [CrossRef]
54. Lechner, J.; Chen, M.; Hogg, R.E.; Toth, L.; Silvestri, G.; Chakravarthy, U.; Xu, H. Higher plasma levels of complement C3a, C4a and C5a increase the risk of subretinal fibrosis in neovascular age-related macular degeneration. *Immun. Ageing* **2016**, *13*, 4. [CrossRef]
55. Katschke, K.J., Jr.; Xi, H.; Cox, C.; Truong, T.; Malato, Y.; Lee, W.P.; McKenzie, B.; Arceo, R.; Tao, J.; Rangell, L.; et al. Classical and alternative complement activation on photoreceptor outer segments drives monocyte-dependent retinal atrophy. *Sci. Rep.* **2018**, *8*, 7348. [CrossRef]
56. Udsen, M.; Tagmose, C.; Garred, P.; Nissen, M.H.; Faber, C. Complement activation by RPE cells preexposed to TNFα and IFNγ. *Exp. Eye Res.* **2022**, *218*, 108982. [CrossRef]
57. Khandhadia, S.; Cipriani, V.; Yates, J.R.; Lotery, A.J. Age-related macular degeneration and the complement system. *Immunobiology* **2012**, *217*, 127–146. [CrossRef]
58. Nishiguchi, K.M.; Yasuma, T.R.; Tomida, D.; Nakamura, M.; Ishikawa, K.; Kikuchi, M.; Ohmi, Y.; Niwa, T.; Hamajima, N.; Furukawa, K.; et al. C9-R95X polymorphism in patients with neovascular age-related macular degeneration. *Investig. Ophthalmol. Vis. Sci.* **2012**, *53*, 508–512. [CrossRef]
59. Rohrer, B.; Guo, Y.; Kunchithapautham, K.; Gilkeson, G.S. Eliminating complement factor D reduces photoreceptor susceptibility to light-induced damage. *Investig. Ophthalmol. Vis. Sci.* **2007**, *48*, 5282–5289. [CrossRef]
60. Qin, S.; Dong, N.; Yang, M.; Wang, J.; Feng, X.; Wang, Y. Complement Inhibitors in Age-Related Macular Degeneration: A Potential Therapeutic Option. *J. Immunol. Res.* **2021**, *2021*, 9945725. [CrossRef]
61. Fong, D. Spotlight on ANX007—Inhibition of C1q for GA. In Proceedings of the 2nd Dry AMD Therapeutic Development Summit, Online, 19–20 October 2021.
62. Jiao, H.; Rutar, M.; Fernando, N.; Yednock, T.; Sankaranarayanan, S.; Aggio-Bruce, R.; Provis, J.; Natoli, R. Subretinal macrophages produce classical complement activator C1q leading to the progression of focal retinal degeneration. *Mol. Neurodegener* **2018**, *13*, 45. [CrossRef]
63. Hughes, S.; Gumas, J.; Lee, R.; Rumano, M.; Berger, N.; Gautam, A.K.; Sfyroera, G.; Chan, A.L.; Gnanaguru, G.; Connor, K.M.; et al. Prolonged intraocular residence and retinal tissue distribution of a fourth-generation compstatin-based C3 inhibitor in non-human primates. *Clin. Immunol.* **2020**, *214*, 108391. [CrossRef] [PubMed]
64. Kim, B.J.; Hughes, S.; Gumas, J.; Rumano, M.; Yancopoulou, D.; Mastellos, D.; Lambris, J.D. Prolonged Intraocular residence of a fourth generation compstatin complement C3 inhibitor supports its clinical development for geographic atrophy. In Proceedings of the Retina Society, Virtual, 25 August–22 September 2020.
65. Liao, D.S.; Grossi, F.V.; El Mehdi, D.; Gerber, M.R.; Brown, D.M.; Heier, J.S.; Wykoff, C.C.; Singerman, L.J.; Abraham, P.; Grassmann, F.; et al. Complement C3 Inhibitor Pegcetacoplan for Geographic Atrophy Secondary to Age-Related Macular Degeneration: A Randomized Phase 2 Trial. *Ophthalmology* **2020**, *127*, 186–195. [CrossRef] [PubMed]
66. Nittala, M.G.; Metlapally, R.; Ip, M.; Chakravarthy, U.; Holz, F.G.; Staurenghi, G.; Waheed, N.; Velaga, S.B.; Lindenberg, S.; Karamat, A.; et al. Association of Pegcetacoplan With Progression of Incomplete Retinal Pigment Epithelium and Outer Retinal Atrophy in Age-Related Macular Degeneration: A Post Hoc Analysis of the FILLY Randomized Clinical Trial. *JAMA Ophthalmol.* **2022**, *140*, 243–249. [CrossRef] [PubMed]
67. Wykoff, C.C.; Rosenfeld, P.J.; Waheed, N.K.; Singh, R.P.; Ronca, N.; Slakter, J.S.; Staurenghi, G.; Mones, J.; Baumal, C.R.; Saroj, N.; et al. Characterizing New-Onset Exudation in the Randomized Phase 2 FILLY Trial of Complement Inhibitor Pegcetacoplan for Geographic Atrophy. *Ophthalmology* **2021**, *128*, 1325–1336. [CrossRef]
68. Apellis Completes Enrollment in Two Phase 3 Studies of the Targeted C3 Therapy, Pegcetacoplan, in Patients with Geographic Atrophy (GA). Available online: https://investors.apellis.com/news-releases/news-release-details/apellis-completes-enrollment-two-phase-3-studies-targeted-c3 (accessed on 30 May 2022).
69. Goldberg, R.; Heier, J.; Wykoff, C.; Staurenghi, G.; Singh, R.P.; Steinle, N.; Boyer, D.; Mones, J.; Holz, F.G.; Bliss, C.; et al. Efficacy of intravitreal pegcetacoplan in patients with geographic atrophy (GA): 18-month results from the phase 3 OAKS and DERBY studies. In Proceedings of the 2022 ARVO Annual Meeting, Denver, CO, USA, 2 May 2022.

70. Kaushal, S.; Grossi, F.; Francois, C.; Slakter, J.; Group, A.S. Complement C3 inhibitor POT-4: Clinical Safety of Intravitreal Administration. *Investig. Ophthalmol. Vis. Sci.* **2009**, *50*, 5010.
71. Wykoff, C.C.; Hershberger, V.; Eichenbaum, D.; Henry, E.; Younis, H.S.; Chandra, P.; Yuan, N.; Solloway, M.; DePaoli, A. Inhibition of Complement Factor 3 in Geographic Atrophy with NGM621: Phase 1 Dose-Escalation Study Results. *Am. J. Ophthalmol.* **2022**, *235*, 131–142. [CrossRef] [PubMed]
72. Yehoshua, Z.; de Amorim Garcia Filho, C.A.; Nunes, R.P.; Gregori, G.; Penha, F.M.; Moshfeghi, A.A.; Zhang, K.; Sadda, S.; Feuer, W.; Rosenfeld, P.J. Systemic complement inhibition with eculizumab for geographic atrophy in age-related macular degeneration: The COMPLETE study. *Ophthalmology* **2014**, *121*, 693–701. [CrossRef]
73. Filho, C.A.d.A.G.; Yehoshua, Z.; Gregori, G.; Nunes, R.P.; Penha, F.M.; Moshfeghi, A.A.; Zhang, K.; Feuer, W.; Rosenfeld, P.J. Change in Drusen Volume as a Novel Clinical Trial Endpoint for the Study of Complement Inhibition in Age-related Macular Degeneration. *Ophthalmic Surg. Lasers Imaging Retin.* **2014**, *45*, 18–31. [CrossRef]
74. Jaffe, G.J.; Westby, K.; Csaky, K.G.; Mones, J.; Pearlman, J.A.; Patel, S.S.; Joondeph, B.C.; Randolph, J.; Masonson, H.; Rezaei, K.A. C5 Inhibitor Avacincaptad Pegol for Geographic Atrophy Due to Age-Related Macular Degeneration: A Randomized Pivotal Phase 2/3 Trial. *Ophthalmology* **2021**, *128*, 576–586. [CrossRef]
75. Sheth, V. Avacincaptad Pegol, A Novel C5 Inhibitor, Demonstrates a Continued Reduction in the Mean Rate of Geographic Atrophy Growth: 18-Month Results from the GATHER1 Clinical Trial. *Investig. Ophthalmol. Vis. Sci.* **2021**, *62*, 333.
76. Jaffe, G.J.; Sahni, J.; Fauser, S.; Geary, R.S.; Schneider, E.; McCaleb, M. Development of IONIS-FB-LRx to Treat Geographic Atrophy Associated with AMD. *Investig. Ophthalmol. Vis. Sci.* **2020**, *61*, 4305.
77. Katschke, K.J., Jr.; Wu, P.; Ganesan, R.; Kelley, R.F.; Mathieu, M.A.; Hass, P.E.; Murray, J.; Kirchhofer, D.; Wiesmann, C.; van Lookeren Campagne, M. Inhibiting alternative pathway complement activation by targeting the factor D exosite. *J. Biol. Chem* **2012**, *287*, 12886–12892. [CrossRef]
78. Yaspan, B.L.; Williams, D.F.; Holz, F.G.; Regillo, C.D.; Li, Z.; Dressen, A.; van Lookeren Campagne, M.; Le, K.N.; Graham, R.R.; Beres, T.; et al. Targeting factor D of the alternative complement pathway reduces geographic atrophy progression secondary to age-related macular degeneration. *Sci. Transl. Med.* **2017**, *9*, eaaf1443. [CrossRef]
79. Holz, F.G.; Sadda, S.R.; Busbee, B.; Chew, E.Y.; Mitchell, P.; Tufail, A.; Brittain, C.; Ferrara, D.; Gray, S.; Honigberg, L.; et al. Efficacy and Safety of Lampalizumab for Geographic Atrophy Due to Age-Related Macular Degeneration: Chroma and Spectri Phase 3 Randomized Clinical Trials. *JAMA Ophthalmol.* **2018**, *136*, 666–677. [CrossRef]
80. Heier, J.S.; Pieramici, D.; Chakravarthy, U.; Patel, S.S.; Gupta, S.; Lotery, A.; Lad, E.M.; Silverman, D.; Henry, E.C.; Anderesi, M.; et al. Visual Function Decline Resulting from Geographic Atrophy: Results from the Chroma and Spectri Phase 3 Trials. *Ophthalmol. Retin.* **2020**, *4*, 673–688. [CrossRef]
81. Cashman, S.M.; Gracias, J.; Adhi, M.; Kumar-Singh, R. Adenovirus-mediated delivery of Factor H attenuates complement C3 induced pathology in the murine retina: A potential gene therapy for age-related macular degeneration. *J. Gene Med.* **2015**, *17*, 229–243. [CrossRef]
82. Cabral de Guimaraes, T.A.; Daich Varela, M.; Georgiou, M.; Michaelides, M. Treatments for dry age-related macular degeneration: Therapeutic avenues, clinical trials and future directions. *Br J. Ophthalmol.* **2022**, *106*, 297–304. [CrossRef]
83. Khanani, A.M.; Maturi, R.K.; Bagheri, N.; Bakall, B.; Boyer, D.S.; Couvillion, S.S.; Dhoot, D.S.; Holekamp, N.M.; Jamal, K.N.; Marcus, D.M.; et al. A Phase I, Single Ascending Dose Study of GEM103 (Recombinant Human Complement Factor H) in Patients with Geographic Atrophy. *Ophthalmol. Sci.* **2022**, *2*, 100154. [CrossRef]
84. Gemini Therapeutics Announces Initial Data from Its Ongoing Phase 2a Study of GEM103 in Patients with Geographic Atrophy Secondary to Dry Age-Related Macular Degeneration. 2021. Available online: https://investors.geminitherapeutics.com/news/news-details/2021/Gemini-Therapeutics-Announces-Initial-Data-From-Its-Ongoing-Phase-2a-Study-of-GEM103-in-Patients-With-Geographic-Atrophy-Secondary-to-Dry-Age-Related-Macular-Degeneration/default.aspx (accessed on 1 June 2022).
85. Gemini Therapeutics Provides GEM103 Program Update. 2022. Available online: https://investors.geminitherapeutics.com/news/news-details/2022/Gemini-Therapeutics-Provides-GEM103-Program-Update/default.aspx (accessed on 30 May 2022).
86. Dreismann, A.K.; McClements, M.E.; Barnard, A.R.; Orhan, E.; Hughes, J.P.; Lachmann, P.J.; MacLaren, R.E. Functional expression of complement factor I following AAV-mediated gene delivery in the retina of mice and human cells. *Gene* **2021**, *28*, 265–276. [CrossRef]
87. Nielsen, J.S. Preliminary Results from a First-in-Human Phase I/II Gene Therapy Trial (FOCUS) of Subretinally Delivered GT005, an Investigational AAV2 Vector, in Patients with Geographic Atrophy Secondary to Age-Related Macular Degeneration. In Proceedings of the ARVO 2022 Annual Meeting, Denver, CO, USA, 1–4 May 2022.
88. Kumar-Singh, R. The role of complement membrane attack complex in dry and wet AMD—From hypothesis to clinical trials. *Exp. Eye Res.* **2019**, *184*, 266–277. [CrossRef] [PubMed]
89. Cashman, S.M.; Ramo, K.; Kumar-Singh, R. A non membrane-targeted human soluble CD59 attenuates choroidal neovascularization in a model of age related macular degeneration. *PLoS ONE* **2011**, *6*, e19078. [CrossRef] [PubMed]
90. Dugel, P. Clinical Trial Download: Data on a Gene Therapy for Dry and Wet AMD. *Retin. Physician* **2020**, *17*, 16–17.
91. Johnson, L.; Splawski, I.; Baker, L.; Carrion, A.; Nguyen, A.; Twarog, M.; Wang, Y.; Jager, U.; Keating, M.; Dryja, T.P.; et al. Generation and characterization of CLG561: A fully-human, anti-properdin Fab for the treatment of age-related macular degeneration. *Investig. Ophthalmol. Vis. Sci.* **2016**, *57*, 1116.

92. Glazer, L.C.; Williams, J.G.; Gordon, C.M.; Dugel, P.U.; Milton, M.; Valencia, T.; Klein, U.; Kretz, S.; Gedif, K.; Grosskreutz, C.L.; et al. A first in human study of Intravitreal (IVT) CLG561 in Subjects with Advanced Age-Related Macular Degeneration (AMD). *Investig. Ophthalmol. Vis. Sci.* **2016**, *57*, 2672.
93. Abokyi, S.; To, C.-H.; Lam, T.T.; Tse, D.Y. Central Role of Oxidative Stress in Age-Related Macular Degeneration: Evidence from a Review of the Molecular Mechanisms and Animal Models. *Oxidative Med. Cell. Longev.* **2020**, *2020*, 7901270. [CrossRef] [PubMed]
94. Bird, A.C.; Bok, D. Why the macula? *Eye* **2018**, *32*, 858–862. [CrossRef]
95. Rastoin, O.; Pagès, G.; Dufies, M. Experimental Models in Neovascular Age Related Macular Degeneration. *Int. J. Mol. Sci.* **2020**, *21*, 4627. [CrossRef]
96. Malek, G.; Busik, J.; Grant, M.B.; Choudhary, M. Models of retinal diseases and their applicability in drug discovery. *Expert Opin. Drug Discov.* **2018**, *13*, 359–377. [CrossRef]
97. Belgio, B.; Boschetti, F.; Mantero, S. Towards an In Vitro Retinal Model to Study and Develop New Therapies for Age-Related Macular Degeneration. *Bioengineering* **2021**, *8*, 18. [CrossRef]
98. Lin, J.B.; Serghiou, S.; Miller, J.W.; Vavvas, D.G. Systemic Complement Activation Profiles in Nonexudative Age-Related Macular Degeneration: A Systematic Review. *Ophthalmol. Sci.* **2022**, *2*, 100118. [CrossRef]
99. Yates, J.R.; Sepp, T.; Matharu, B.K.; Khan, J.C.; Thurlby, D.A.; Shahid, H.; Clayton, D.G.; Hayward, C.; Morgan, J.; Wright, A.F.; et al. Complement C3 variant and the risk of age-related macular degeneration. *N. Engl. J. Med.* **2007**, *357*, 553–561. [CrossRef]
100. van Leeuwen, E.M.; Emri, E.; Merle, B.M.J.; Colijn, J.M.; Kersten, E.; Cougnard-Gregoire, A.; Dammeier, S.; Meester-Smoor, M.; Pool, F.M.; de Jong, E.K.; et al. A new perspective on lipid research in age-related macular degeneration. *Prog. Retin. Eye Res.* **2018**, *67*, 56–86. [CrossRef]
101. Xu, Q.; Cao, S.; Rajapakse, S.; Matsubara, J.A. Understanding AMD by analogy: Systematic review of lipid-related common pathogenic mechanisms in AMD, AD, AS and GN. *Lipids Health Dis.* **2018**, *17*, 3. [CrossRef]
102. Matsubara, J.A.; Tian, Y.; Cui, J.Z.; Zeglinski, M.R.; Hiroyasu, S.; Turner, C.T.; Granville, D.J. Retinal Distribution and Extracellular Activity of Granzyme B: A Serine Protease That Degrades Retinal Pigment Epithelial Tight Junctions and Extracellular Matrix Proteins. *Front. Immunol.* **2020**, *11*, 574. [CrossRef]
103. Dörschmann, P.; Klettner, A. Fucoidans as Potential Therapeutics for Age-Related Macular Degeneration-Current Evidence from In Vitro Research. *Int. J. Mol. Sci.* **2020**, *21*, 9272. [CrossRef]
104. Sobolewska, B.; Sabsabi, M.; Ziemssen, F. Importance of Treatment Duration: Unmasking Barriers and Discovering the Reasons for Undertreatment of Anti-VEGF Agents in Neovascular Age-Related Macular Degeneration. *Clin. Ophthalmol.* **2021**, *15*, 4317–4326. [CrossRef]
105. Robin, A.L.; Muir, K.W. Medication adherence in patients with ocular hypertension or glaucoma. *Expert Rev. Ophthalmol.* **2019**, *14*, 199–210. [CrossRef]
106. Lynch, C.R.; Kondiah, P.P.D.; Choonara, Y.E.; du Toit, L.C.; Ally, N.; Pillay, V. Hydrogel Biomaterials for Application in Ocular Drug Delivery. *Front. Bioeng. Biotechnol.* **2020**, *8*, 228. [CrossRef]
107. Cooper, R.C.; Yang, H. Hydrogel-based ocular drug delivery systems: Emerging fabrication strategies, applications, and bench-to-bedside manufacturing considerations. *J. Control. Release* **2019**, *306*, 29–39. [CrossRef]
108. Tsai, C.H.; Wang, P.Y.; Lin, I.C.; Huang, H.; Liu, G.S.; Tseng, C.L. Ocular Drug Delivery: Role of Degradable Polymeric Nanocarriers for Ophthalmic Application. *Int. J. Mol. Sci.* **2018**, *19*, 2830. [CrossRef] [PubMed]
109. Costa, J.R.; Silva, N.C.; Sarmento, B.; Pintado, M. Potential chitosan-coated alginate nanoparticles for ocular delivery of daptomycin. *Eur. J. Clin. Microbiol. Infect. Dis.* **2015**, *34*, 1255–1262. [CrossRef] [PubMed]

biomedicines

Tutorial

Methodological Appraisal of Phase 3 Clinical Trials in Geographic Atrophy

Marc Biarnés [1,*], Xavier Garrell-Salat [1,2], Alba Gómez-Benlloch [1,2], Mercè Guarro [1,2], Gabriel Londoño [1,2], Elena López [1], Sergi Ruiz [1], Meritxell Vázquez [1] and Laura Sararols [1,2]

1 OMIQ Research, Carrer de Pedro i Pons 1, 08195 Sant Cugat del Vallès, Spain; xgarrell@omiq.es (X.G.-S.); agomez@omiq.es (A.G.-B.); mguarro@omiq.es (M.G.); glondono@omiq.es (G.L.); elopez@omiq.es (E.L.); sruiz@omiq.es (S.R.); mvazquez@omiq.es (M.V.); lsararols@omiq.es (L.S.)
2 Department of Ophthalmology, Hospital General de Granollers, Av. Francesc Ribas s/n, 08402 Granollers, Spain
* Correspondence: mbiarnes@omiq.es; Tel.: +34-690-79-50-52

Abstract: Geographic atrophy (GA) secondary to age-related macular degeneration is a common cause of blindness worldwide. Given the recent approval of the first therapy for GA, pegcetacoplan, we critically appraise methodological aspects of the phase 3 clinical trials published so far in this disease in relation to their design, analysis and interpretation. We reviewed some of the key attributes of all phase 3 clinical trials in GA available in the main public registry of clinical trials as of 20 May 2023. The topics discussed included types of endpoints, eligibility criteria, *p*-value and effect size, study power and sample size, the intention to treat principle, missing data, consistency of results, efficacy–safety balance and application of results. Five phase 3 clinical trials have reported results, either partially or completely: GATHER1, DERBY/OAKS, CHROMA/SPECTRI, SEATTLE and GATE. Although there are many similarities between these trials in terms of endpoints or broad eligibility criteria, they differ in several aspects (metric of the primary endpoint, sample size, type of adverse events, etc.) that can influence the results, which are discussed. Readers should understand key methodological aspects of clinical trials to improve their interpretation. On the other hand, authors should adhere to clinical trial reporting guidelines to communicate what was done and how it was done.

Keywords: age-related macular degeneration; clinical trials; endpoints; geographic atrophy; statistics

Citation: Biarnés, M.; Garrell-Salat, X.; Gómez-Benlloch, A.; Guarro, M.; Londoño, G.; López, E.; Ruiz, S.; Vázquez, M.; Sararols, L. Methodological Appraisal of Phase 3 Clinical Trials in Geographic Atrophy. *Biomedicines* **2023**, *11*, 1548. https://doi.org/10.3390/biomedicines11061548

Academic Editor: Enzo Maria Vingolo

Received: 12 April 2023
Revised: 20 May 2023
Accepted: 24 May 2023
Published: 26 May 2023

1. Introduction

Geographic atrophy (GA) is the advanced form of dry age-related macular degeneration (AMD; Figure 1). It is characterized by the progressive loss of photoreceptors, retinal pigment epithelium (RPE) and choriocapillaris [1]. As such, areas affected by GA correspond to absolute scotomata [2], which will affect best-corrected visual acuity (BCVA) if the fovea is involved [3]. Nonetheless, even if atrophy surrounds it (the so-called foveal sparing), patients experience difficulties in reading, driving and recognizing faces, which impairs their quality of life [4,5]. The estimated worldwide prevalence of GA is 5 million, which will probably grow in the next decades due to increases in life expectancy. No treatment was available for this condition until 17 February 2023, when Apellis Pharmaceuticals announced that the Food and Drug Administration (FDA) had approved Syfovre® (intravitreal pegcetacoplan 15 mg) based on a slower progression of atrophy in the DERBY and OAKS phase 3 randomized controlled clinical trials.

We reviewed published phase 2/3 and 3 clinical trials in GA to highlight some aspects of their design, analysis, and interpretation (Table 1). Phase 1 and 2 trials have specific aims (dose finding, study of pharmacokinetics, etc.), and will not be addressed here. The purpose of this manuscript is to critically appraise key issues in clinical trials to aid in their interpretation using examples from the field of GA.

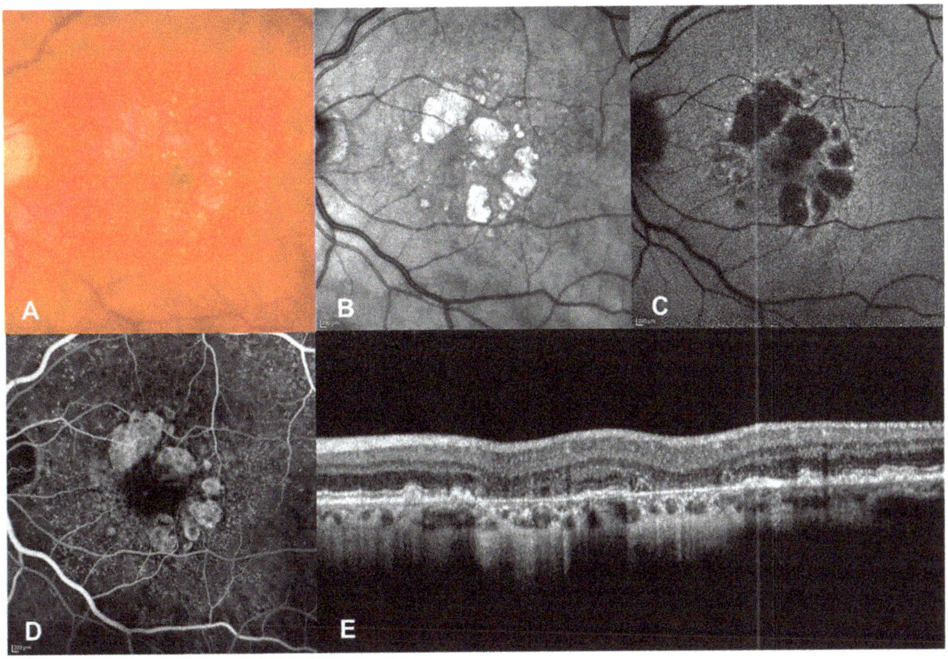

Figure 1. Geographic atrophy on multimodal imaging. (**A**) color fundus photography; (**B**) infrared; (**C**) fundus autofluorescence; (**D**) fluorescein angiography; (**E**) extrafoveal spectral domain optical coherence tomography. The perifoveal, multifocal areas of RPE atrophy are hyperreflective in (**B**), hypoautofluorescent in (**C**), seen as window defects in (**D**), and as RPE and outer nuclear layer loss with secondary retinal thinning and choroidal hypertransmission in (**E**). Scattered drusen are also present. RPE: retinal pigment epithelium.

Table 1. Topics discussed.

Design	Analysis	Interpretation
Primary, secondary, and composite endpoints Eligibility criteria	*p*-value and effect size Study power/sample size Intention to treat and treatment adherence Missing data Consistency of results	Efficacy–safety balance Application of results to my patients

Phase 3 Clinical Trials in Geographic Atrophy

There have been many trials on GA conducted for the past 15 years, but only a few reached phase 3 and have been posted in www.clinicaltrials.gov (accessed on 11 April 2023; Table 2). Only those with available complete or preliminary results released on the primary endpoint will be discussed: GATHER1 (NCT02686658) [6], DERBY (NCT03525600) and OAKS (NCT03525613, the preliminary results of which have been reported in different Meetings between 2021 and 2023 and are available at the Apellis website [7]), CHROMA (NCT02247479) and SPECTRI (NCT02247531) [8], SEATTLE (NCT01802866) [9] and GATE (NCT00890097) [10].

Table 2. Summary of phase 2/3 or 3 clinical trials in GA secondary to age-related macular degeneration.

Trial	Sponsor	Patients (Total *n*)	Drug, Route	MoA	Primary Endpoint	Results
SAGA, Sep/23	Alkeus	≥60 years At least one eye with GA (ongoing; planned *n* = 300)	ALK-001, oral	Slowing vitamin A dimerization	Difference in growth rate	Study ongoing
GATHER1/2, [6] Jul/23	IVERIC bio	≥50 years Non-foveal GA, but atrophy ≤ 1500 μm from foveal center (*n* = 286)	Avacincaptad pegol (Zimura®), IVI	Complement regulation (anti-C5)	Difference in growth rate at 12 months after the square root transformation *	(% reduction in treated vs. sham) 2 mg: 27.4% (p = 0.0072) 4 mg: 27.8% (p = 0.0051)
DERBY-OAKS, [7] Jan/23	Apellis	≥60 years GA area ≥ 2.5 and ≤17.5 mm^2 (*n* = 1.258)	Pegcetacoplan (Syfovre®), IVI	Complement regulation (anti-C3)	Difference in growth rate at 12 months	(% reduction in treated vs. sham) DERBY: Monthly: 12% (p = 0.0528) EOM: 11% (0.0750) OAKS: Monthly: 22% (p = 0.0003) EOM: 16% (p = 0.0052)
TOGA, Nov/20	MEDARVA Foundation	≥55 years GA ≥ 0.5 and ≤7 disc areas (*n* = 286)	Doxycycline (Oracea®), oral	Antiinflammatory	Difference in growth rate at months 6 and 30	Not published
CHROMA-SPECTRI, [8] Jan/18	Roche	≥50 years Well demarcated GA area (*n* = 1.881)	Lampalizumab, IVI	Complement regulation (anti-Factor D)	Difference in mean change in GA area at week 48	(lampalizumab minus sham) CHROMA: q4w: −0.02 mm^2 (p = 0.80), q6w: +0.05 mm^2 (p = 0.59); SPECTRI: q4w: +0.16 mm^2 (p = 0.048 favoring sham), q6w: +0.09 mm^2 (p = 0.27)
SEATTLE, [9] May/16	Acucela Inc.	≥55 years Clinical diagnosis of GA (*n* = 508)	Emixustat hydrochloride (ACU-4429), oral	Visual cycle modulation (anti-RPE65)	Difference in growth rate over 24 months	Treated: 1.69 to 1.84 mm^2/y Sham-treated: 1.69 mm^2/y ($p \geq 0.81$)
GATE, [10] May/12	Alcon	≥55 years GA area ≥ 1.25 and ≤20 mm^2 (*n* = 768)	AL-8309B (tandospirone), topical ocular	Neuroprotective (5-HT1A agonist)	Difference in mean annualized lesion growth rate up to 30 months	AL-8309B 1.0%: 1.73 mm^2/y AL-8309B 1.75%: 1.76 mm^2/y Vehicle: 1.71 mm^2/y

Date under trial name is its estimated study completion date. Available on https://clinicaltrials.gov/, accessed on 20 May 2023. BCVA: best-corrected visual acuity; EOM: every other month; GA: geographic atrophy; IVI: intravitreal injection; MoA: mechanism of action; q4w: treatment every 4 weeks; q6w: treatment every 6 weeks; y: year. * This is based on subtracting the square root of the area of atrophy at the end of study from the square root of the area of atrophy at baseline divided by the time between visits in years to determine GA growth.

Briefly, in GATHER1, intravitreal injections of 2 mg or 4 mg avacincaptad pegol (Zimura®, IVERIC Bio Inc., Cranbury, NJ, USA), a C5 inhibitor, were administered monthly [6]. In DERBY/OAKS, pegcetacoplan 15 mg (Syfovre®, Apellis Pharmaceuticals, Waltham, MA, USA), a C3 inhibitor upstream of C5, was administered monthly or every other month (EOM) by intravitreal injections [7]. CHROMA and SPECTRI used 10 mg lampalizumab

(F. Hoffmann-La Roche Ltd., Basel, Switzerland), a complement factor D inhibitor, delivered by intravitreal injection every 4 or 6 weeks [8]. In SEATTLE, emixustat hydrochloride (formerly ACU-4429, Acucela, Inc., Seattle, WA, USA), an inhibitor of the visual cycle isomerhydrolase RPE65, was administered orally once daily at 2.5 mg, 5 mg or 10 mg doses [9]. Finally, in GATE AL-8309B (tandospirone; Alcon, Fort Worth, TX, USA), a neuroprotective 5-HT1A agonist, was topically administered twice daily at concentrations of 1.0% or 1.75% [10]. All trials compared these drugs to sham treatment.

2. Design

2.1. Primary, Secondary and Composite Endpoints

The primary endpoint is the main result measured at the end of the study. A positive result in the primary endpoint is a prerequisite to accept the new intervention as effective, and it is also used as the reference for sample size calculations. Secondary endpoints evaluate additional effects of therapy. Composite endpoints combine two or more outcomes in a single endpoint to increase the number of events and decrease sample size requirements, trial length or costs.

All included phase 3 trials used the difference in the anatomical outcome between study arms, GA growth, as the primary endpoint. Fundus autofluorescence (FAF) was the imaging method used in all studies to measure the area of atrophy, but in future it may be replaced by optical coherence tomography (OCT), because this provides additional information [11], like outer nuclear layer thickness and early fluid detection. Of note, GA is a surrogate endpoint, a measurement used as a substitute of a clinically meaningful endpoint that measures directly how a patient feels, functions or survives [12]. Nonetheless, it has already been accepted by the FDA as a primary endpoint [13].

All trials used mixed-effects models to account for the repeated measurements of atrophy taken in each participant throughout the study, and all but one measured the primary endpoint in mm^2/year (area). Instead, GATHER1 used square root transformation [14] to express the results in mm/year (linear growth), which also decreases the association between baseline area of atrophy and growth. While randomization should achieve balance between study arms in baseline area of atrophy and other potential predictors of growth (location of atrophy, multifocality, etc.), imbalances may occur by chance. To decrease this source of potential residual confounding, CHROMA/SPECTRI, DERBY/OAKS and SEATTLE included different baseline covariates in the mixed-effects models, providing an adjusted estimate of growth.

Secondary endpoints have included mainly functional outcomes, like changes in BCVA in standard and low-luminance conditions, low-luminance deficit, macular sensitivity on microperimetry, reading speed or patient-reported outcome measures (PROM) via questionnaires of quality of life related to vision [15]. Arguably, functional endpoints are more relevant to patients and are preferrred by regulatory agencies [13]. However, they have higher intra-subject variability than structural outcomes, and current tests show a slow deterioration in the typical duration of a phase 3 trial (1.5–2 years), which makes detection of changes on visual function caused by therapy difficult.

No composite endpoints have been used so far. A combination of functional endpoints may increase the sensitivity to detect visual changes in GA. Together with the development of new functional tests, this area deserves further research.

The time to a given untoward outcome is common in other fields, but rare in GA trials (Figure 2A). These endpoints are usually analyzed with time-to-event modeling, like proportional hazards or parametric models, and plotted with Kaplan–Meier graphs (Figure 2B). The results are usually summarized as hazard ratios, which express the relative risk of experiencing the event in treated vs. control arms, given that event has not already occurred up to that point. Of note, the effect of therapy could also be expressed by the added time free of the event in the treatment arm using accelerated failure time models, [16] which may be relevant in, for example, incident subfoveal atrophy (Figure 2B). This metric is also more intuitive for the patient.

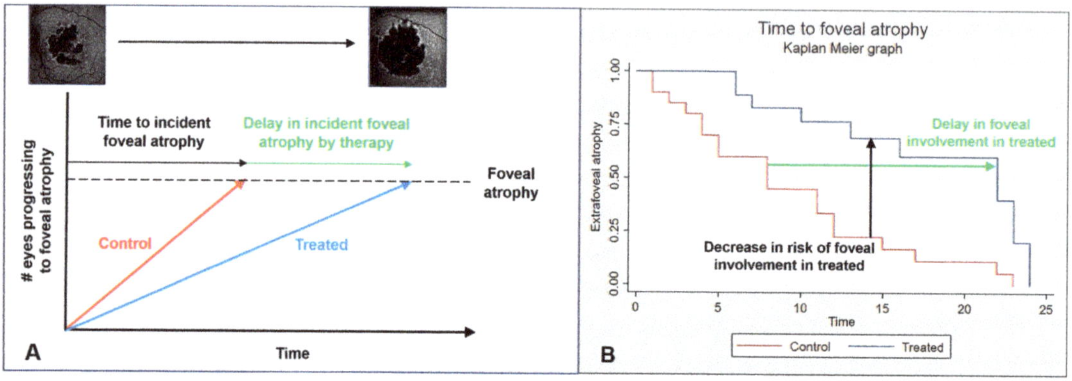

Figure 2. Alternative endpoints. (**A**) In extrafoveal lesions, time to, say, subfoveal atrophy should be delayed in treated (blue line) vs. untreated (red line) eyes. (**B**) From conventional Kaplan–Meier curves, one can derive the hazard ratio, the risk of progression to subfoveal atrophy by comparing treated vs. untreated eyes (vertical black line); alternatively, one can estimate the time gained without subfoveal atrophy from therapy (horitzontal green line), which is possibly a more intuitive metric.

2.2. Eligibility Criteria

Inclusion and exclusion criteria define the set of characteristics that make patients eligible to participate in the trial. Eligibility criteria are the basis used to determine the extent to which the study results can be applied to other patients (i.e., generalizability).

The percentage of screening failures and the reasons for exclusion provide important data in this regard. The percentage of screening failure was 65% in GATHER1, 59% in CHROMA/SPECTRI and 47.5% in SEATTLE, while data are not yet available for DERBY/OAKS and were not available for GATE. When reported, the main reason for failure was ineligible GA area, suggesting that clear definitions and education on measuring the area of atrophy can minimize patient inconvenience and trial costs.

Regarding specific criteria, GATHER1 and CHROMA/SPECTRI included patients 50 years or older with GA (Table 2). This is surprising, considering that atrophy takes years to ensue, and therefore some patients could have had atrophy developing before their fifties, making alternative diagnosis more likely. In one study, more than 5% of patients with a diagnosis of dry AMD had variants in genes known to cause an inherited retinal disease when genetic testing was conducted [17], suggesting that selection of older patients is prudent to decrease the chance of mistakenly including non-GA conditions.

Careful thought should be given to inclusion of eyes with subfoveal atrophy. The advantage of excluding these eyes is that extrafoveal GA lesions progress faster [18] and the effects of therapy can therefore be established in less time and with a smaller sample. Additionally, differences in prevention of BCVA loss by incident subfoveal atrophy can presumably be established, as long as a large enough number of eyes experience this event (Figure 2). The drawbacks include a longer recruitment period and application of results to a subgroup of all patients with GA, probably less than 50% [19]. GATHER1 is the only phase 3 trial that excluded subfoveal lesions, and its results seem to support this strategy because avacincaptad pegol decreased growth by almost 30% at 12 months [6]. Similar results have been found in subgroup analyses in DERBY/OAKS with pegacetacoplan [7]. Including both locations of atrophy implicitly assumes that the same disease pathway operates in subfoveal and extrafoveal GA, which is a matter of debate [20,21].

3. Analysis

3.1. p-Value and Effect Size

The *p*-value is one of the most widely used and poorly understood statistics. It is the probability of finding a result as extreme as that obtained or greater if the null hypothesis is true. As such, it is a conditional probability: conditional on the null hypothesis being true. Although they are related, the p-value does not convey direct information on the effect size, which is key for medical practice. For this, one should look at absolute (difference) or even relative (ratio) metrics between trial arms and their confidence intervals (CI), which provide information regarding the likely magnitude of the effect of therapy.

In CHROMA/SPECTRI, SEATTLE and GATE, the mean progression of GA was the same or slightly larger in the treated group, and therefore p-values were very high or not reported at all. CHROMA/SPECTRI pre-specified a subgroup analysis in carriers of the complement factor I (CFI) risk allele, based on results of the small (n = 129) phase 2 MAHALO [22]. In MAHALO, CFI was one of four variants in the complement (with CFH, C2/CFB and C3) preselected for subgroup analyses. The researchers found a 49% faster growth in CFI carriers than in noncarriers in the sham group, and a 44% reduction in GA growth in CFI+ patients treated with lampalizumab as compared to sham. However, in the phase 3 trials, CHROMA/SPECTRI treatment did not slow GA progression, either overall or in CFI carriers [8]. Chance may explain the MAHALO findings, since the faster progression in CFI carriers could not be replicated in other studies [23], its biological basis is uncertain, and the analyses were not adjusted for multiple comparisons. Additionally, 26.4% of treated eyes (23/87) discontinued the study, and the main imputation method used to complete the missing data (the last observation carried forward) can induce bias [24]. Those results underscore the limitations of subgroup analyses in small samples.

On the other hand, GATHER1 and DERBY/OAKS reported positive results. As of April 2023, only pegcetacoplan has been FDA approved for GA, because results of the second, pivotal, trials required for approval of avacincaptad pegol, GATHER2, are not available yet. In GATHER1, treatment reduced progression of GA at 12 months from 0.40 mm (sham) to 0.29 mm (avacincaptad pegol 2 mg), a 27.4% reduction that was statistically significant at a p-value = 0.0072. The growth in the control arm was similar to that reported in untreated extrafoveal lesions in other studies [25], the effect was clinically relevant, the p-value was well below 0.05, and similar results were observed with the 4 mg dose (with no apparent dose–response relationship), which together provide convincing evidence of a real effect of therapy.

The 12-month results of DERBY/OAKS were difficult to interpret. OAKS revealed a reduction in growth compared to sham of 22% in the monthly and 16% in the EOM arms that was statistically significant (p = 0.0003 and 0.0052, respectively). However, DERBY lacked statistical significance, with a reduction of 12% in the monthly (p = 0.0528) and 11% in the EOM (p = 0.0750) arms [7]. How to proceed when the result of one of the pivotal trials is positive and the other is not? The results of both studies suggested a positive effect of the drug on GA progression, a dose–response relationship and the prespecified combined analysis showed a 17% reduction in the monthly (p < 0.0001) and a 14% reduction in the EOM (p = 0.0012) arms, with p-values well below 0.05. Additionally, growth was slower in the treated than in the untreated fellow eyes in the same patients (while growth in the sham treated eyes was similar to that of untreated fellow eyes in the same patients), and the results were in line with those from the phase 2 trial FILLY [26]. This supports a real effect of therapy on growth despite borderline results in OAKS and highlights the importance of looking at the whole evidence when results of a particular study seem inconclusive, which will be also discussed later on.

A related but different matter is effect size, used to assess if the difference is clinically meaningful. The p-value says if the effect of the therapy is probably real, while effect size helps to determine if it matters to patients. The point estimates (mean effects) of C3 or C5 inhibitors on GA progression at 12 months range from approximately −10% to −30%, depending on the drug, frequency of administration, GA location, etc. They represent the most likely value, but 95% CI provides a range of likely values in which the true effect may lie. Narrow 95% CI are desirable because they provide more precise estimates of the effect. For example, GATHER1 reported a mean difference between avacincaptad pegol 4 mg and sham of 0.124 mm/year with 95% CI between 0.038 and 0.209. These 95% CI were wide, reflecting the uncertainty caused by the relatively small number of patients. The drug effect on decreasing GA progression could be as small as 0.038 mm/year or as large as 0.209 mm/year, and it is probably real ($p = 0.0072$): if the drug is truly no different from sham, the chance of finding this difference is 0.72%.

The standardized mean difference (SMD) is used for measuring effect size in a common scale. It is the difference in mean change between treated and placebo arms divided by the pooled standard deviation [27]. SMD of 0.2, 0.5 and 0.8 are regarded as small, medium and large, respectively. Using the published results in GATHER1, the SMD of avacincaptad pegol vs. sham at 12 months was 0.15 (2 mg) and 0.18 (4 mg). Other metrics have been discussed elsewhere [27].

3.2. *Study Power/Sample Size*

The power of a study is the probability of showing a difference between treatments when it truly exists. For a given level of significance or false positive rate (typically set at a two-sided $\alpha = 0.05$ or 5% in superiority trials), power increases when the difference in the primary endpoint between study arms increases, when the variability of the primary endpoint is low and when sample size is large. In addition, an allocation ratio (the distribution of trial participants to treatment or control arms) of 1:1 requires a smaller sample than other ratios (2:1, 3:1, etc.), but this comes at the expense of being less attractive to patients in placebo-controlled trials, and a decreased ability to detect adverse events of the new therapy.

A summary of the assumptions used to calculate the sample size is provided in Table 3. Assumptions for sample size calculations in GATHER1 with avacincaptad pegol were derived from the phase 2 trial FILLY with pegcetacoplan [6]. Sample size assumptions from DERBY/OAKS were not yet available, and GATE did not report them in its main publication. The total sample size ranged from 286 in GATHER1 to 975 in SPECTRI, a ratio of 3.4. The estimated annual growth rates in the control group differed modestly between trials, but effect sizes with reductions of 20–30% of GA growth and estimated yearly discontinuations near 15% were common.

Table 3. Summary of assumptions used to calculate sample size.

	GATHER1	CHROMA/SPECTRI *	SEATTLE
Power (1-β), %	90	88	80
Approximate α (unadjusted)	0.025 **	0.0495	0.05
Growth in the control group (SD), mm^2/y	0.33 (0.20) ***	2.08 (1.53)	1.75 (1.20)
Difference between trial arms (SD), mm^2/y	0.10 (0.20) ***	0.42 (1.53)	0.56 (1.20)
Randomization ratio	1:2:2	2:1:2:1	1:1:1:1
Expected yearly drop-out rate, %	15–20	15	16.7

* Overall patient population; ** one-sided test; *** mm/year; SD, standard deviation.

3.3. *Intention-to-Treat (ITT) Principle*

The ITT analysis specifies that patients (or eyes) are analyzed in the treatment group in which they were originally randomized, regardless of the treatment they eventually received [28]. That means that if someone was randomized to receive an intravitreal injection, but they received the sham procedure, they will be analyzed in the active treatment group. Simply stated, it is the indication or intention to treat a subject with a given treatment (defined by randomization) that matters. This is the standard approach to data analysis in phase 3 trials, because it includes all patients, preserves the benefits of randomization (balancing known and unknown factors that may affect response to treatment between groups), and provides a conservative estimate of the effect of treatment, which is closer to what would be observed if it were administered to the population as a whole. The *per-protocol* (PP) analysis population includes only cases fully adherent to the protocol [28];

as such, PP shows the maximal effect of therapy, i.e., under optimal compliance. Finally, the *as-treated* population analyzes patients according to treatment actually received [28]. Unfortunately, PP and *as-treated* analyses can induce selection bias and confounding.

Except for GALE, where this information was not provided, all trials acknowledged following the ITT principle. In some trials, a *modified* ITT analysis was used, meaning that patients were included in the ITT analysis if they met certain prerequisites, like having received at least one intravitreal injection and having a post-treatment visit available. However, a modified ITT may mean different things in different trials, and the reader must pay attention to how it is defined. In SEATTLE and GALE treatment was self-administered for 24 months. In this scenario, incomplete adherence to treatment or crossovers are a concern, and ITT becomes important to minimize bias. In future trials with self-administered therapies, it would be important to monitor non-compliance and compare the results from ITT and PP analyses. In the other trials, therapy was administered by medical personnel, and thus compliance is known, although non-adherence, crossovers and missing data could also compromise results.

3.4. Missing Data

Pivotal phase 3 trials in ophthalmology usually follow several hundred patients for 18–24 months. Therefore, missing data is unavoidable. This has two consequences: (1) loss of power (in superiority trials, it is more difficult to show a beneficial effect of therapy, even if it truly exists); and (2) biased results if those who discontinue are different from those who remain in the study or if they withdraw for reasons related to the therapy. Keeping a high visit rate is a challenge, particularly for studies ongoing in the 2020–2022 period (SAGA, GATHER1/2 and DERBY/OAKS) due to the COVID-19 pandemic. Minimizing missing data during the study is paramount, and when this occurs, several statistical methods can be used to impute values and try to decrease its impact. These range from the simple last-observation-carried-forward, in which a missing follow-up value is replaced by the previously observed value, to sophisticated multiple imputation procedures, where an average of the imputed values from multiple imputed datasets is estimated, while acknowledging the uncertainties in the estimated value.

Important visits at which every effort should be made to collect all data are baseline, the visit where the primary endpoint is measured, and the end-of-study visit. Missing data can also affect exams within a visit; for example, a missing FAF image. Discontinuation was very variable between studies and ranged from 7.9% in CHROMA/SPECTRI to 37.0% in SEATTLE. In SEATTLE, the main reason for this was incident adverse events, which generally increased with increasing drug dose. In DERBY/OAKS, 11.4% of patients discontinued the study before month 12, while the percent of missed injections was 11.9% (roughly half of them for COVID-19-related reasons). On the other hand, 29.9% of participants in GALE did not finish the study, but the percentage was similar between study arms. Even in this situation, bias can occur because patients leaving the study may be different to those who remain for reasons related to the treatment. Note that patients who discontinue are included in the ITT/modified ITT analyses, sometimes using imputation methods, and sensitivity analyses may be used to evaluate the robustness of the results under different assumptions [22].

3.5. Consistency of Results

Although for regulatory purposes, the key results are those from pivotal phase 3 trials, all available evidence should be taken into account to determine the effects of a given therapy. This includes data from phase 2 trials or, rarely, external evidence in cases in which the therapy is already approved for another indication and tested for the condition of interest. Some of the methods that can be used to quantitatively integrate this information include meta-analyses and Bayesian approaches.

A meta-analysis combines the results of many studies on the same disease and endpoint to provide the most precise estimate of the effect of therapy [29]. Essentially, it is a weighted average of the effects of therapy across studies. A cumulative meta-analysis performs successive meta-analyses, one each time a new study is added, which makes it possible to determine the effect of the most recent study on the overall estimate. This would have been useful in the case of pegcetacoplan, where the primary endpoint results in OAKS were statistically significant, but those of DERBY were not. An estimated cumulative meta-analysis of the phase 2 FILLY [26], and the phase 3 OAKS and DERBY shows that, while the results of the latter attenuated the benefits of the drug, they remained statistically significant (Figure 3).

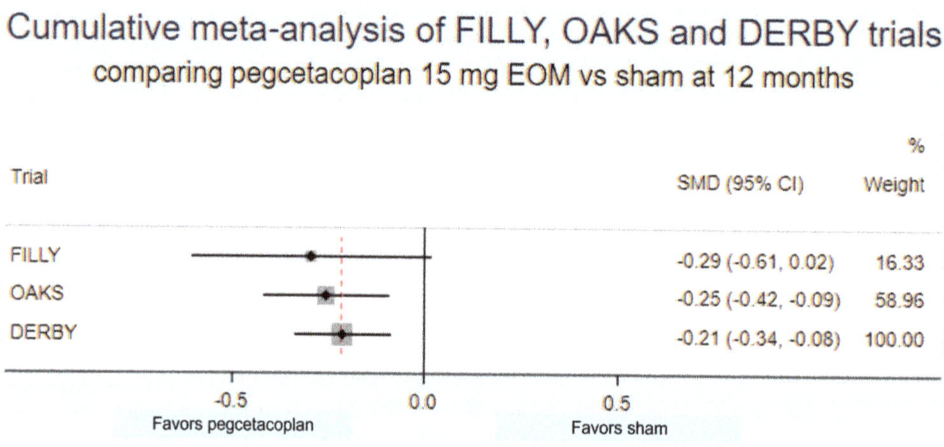

Figure 3. Estimated cumulative meta-analysis of the effects of pegcetacoplan 15 mg every other month on the growth of geographic atrophy at 12 months in the FILLY (phase 2), OAKS and DERBY (phases 3) clinical trials. A modest but real effect is shown when considering all results. CI: confidence interval; SMD: standardized mean difference.

Bayesian methods [30] are less well known and are rarely used in ophthalmology trials. These approaches require that one expresses a plausible distribution of the effects of therapy before the current study based on what is known up to that point (the so-called *prior*), for example the results of a phase 2 trial. Then, one uses Bayes's theorem to update that prior belief with data from recent evidence (the phase 3 trials) to provide a posterior distribution, or a range of likely values were the true effect of therapy lies.

4. Interpretation

4.1. *Efficacy–Safety Balance*

Recent episodes of intraocular inflammation with an otherwise effective antiangiogenic drug in patients with exudative AMD (eAMD) [31] underscore the importance of evaluating both efficacy and safety to establish the effects of an intervention. Table 4 shows a summary of the most commonly observed side effects in the active arm of published GA phase 3 trials. It should be noted, however, that these events were not necessarily more frequent in treated than in placebo arms. We focus the rest of this section on studies in which at least one trial has reported positive results.

The combined results of OAKS and DERBY at 24 months showed an increased rate of new-onset eAMD in treated eyes. There was a dose–response relationship that roughly doubled the risk of eAMD from sham to EOM and from EOM to monthly dosing: 3.1%, 6.7% and 12.2%, respectively. The rates were slightly lower in patients without fellow eye eAMD. The percentage of eAMD cases in the avacincaptad pegol arm was also larger than the sham-treated group in the small GATHER1 trial, suggesting that these drugs do increase exudative events. Cumulative events of intraocular inflammation at 24 months were 2.1% in EOM and 3.8% with monthly dosing, with no reports of occlusive or nonocclusive vasculitis. The rate of endophthalmitis was approximately 1:3000 injections (similar to that reported with conventional antiangiogenic therapy) [32], which is reassuring, considering that these therapies downregulate the innate immune system. Simultaneous treatment of GA with pegcetacoplan and eAMD with antiangiogenic therapy was allowed in DERBY and OAKS, and preliminary results seem to suggest that combined therapy is safe and efficacious. Nonetheless, the increased risk of short-term decrease in BCVA from eAMD must be discussed with the patient when considering treatment with pegcetacoplan for the prevention of long-term vision loss.

Table 4. List of the most common adverse events reported in published phase 3 clinical trials in geographic atrophy in the study treatment arms of each study. For trials with more than one active treatment arm, the results from the two dosing regimens or drug concentrations are pooled.

	GATHER1	DERBY/OAKS *	CHROMA/SPECTRI	SEATTLE	GATE
Ocular	Conj. hemo. (24.7%) eAMD (9.3%) Increased IOP (6.7%)	eAMD (9.4%) ** IOI (3%) Endophthalmitis (1.2%)	Conj. hemo. (29.6%) Increased IOP (11.3%) Eye pain (10%)	Delayed DA (55.4%) Chromatopsia (17.6%) Visual impairment (15.4%)	Reduced VA (32.1%) Eye irritation (9.1%) Cataract (8.1%)
Non-ocular	UTI (9.3%) Fall (6.7%) Nasoph. (6%)	Not reported	URTI (8.6%) Fall 8.6%) Bronchitis (5.5%)	Fall (9.7%) Nasoph. (9.2%) Hypertension (9.6%)	Not reported

Conj. hemo.: conjunctival hemorrhage; DA: dark adaptation; eAMD: exudative age-related macular degeneration; IOI: intraocular inflammation; IOP: intraocular pressure; nasoph: nasopharyngitis; URTI: upper respiratory tract infection; UTI: urinary tract infection; VA: visual acuity. * Pending detailed reporting; ** As determined by the investigator.

4.2. Application of Results to My Patients

It is essential to determine whether the results of the trial can be applied to my patients, because the intervention may be influenced by demographic, biological or socioeconomic factors. A good place to start is reviewing "Table 1" of the trial manuscript, which will usually describe the characteristics of the participants. Other considerations also apply, some of which are discussed here in the context of pegcetacoplan, the only approved therapy as of 11 April 2023.

Since many conditions mimic GA secondary to AMD [33], a first step is confirming that the patient does indeed present this disease. Many disorders can cause macular RPE atrophy, but in none of them has C3 upregulation been demonstrated to play a pathogenic role, and the effect of pegcetacoplan is unknown in these cases. The efficacy and safety of the drug in patients below 60, in non-Caucasians (who represented less than 10% of participants in the phase 3 trials) and with different complement-related genetic variants need to be studied and considered carefully. Results were better in extrafoveal than subfoveal lesions, although the p-values for interaction between treatment and atrophy location are not known by the authors. In terms of safety, the diagnosis in the fellow eye may play a role on the risk of incident eAMD in pegcetacoplan-treated eyes, and the added burden of compliance with two intravitreal drugs may jeopardize results.

Another issue to consider will be adherence to treatment, which is a common reason of poor translation of results from clinical trials [34]. This will depend on both patients and practitioners. The costs of therapy and the organization of healthcare services, public and private, will be determinants of adherence. The open-label extension GALE (NCT04770545), phase 4 and real-world evidence studies will inform the impact of compliance on drug effectiveness.

5. Discussion

Approval of medical therapies is based on sound, rigorous clinical trials. We reviewed some points related to their design, analysis and interpretation, and additional information on similar topics can be found elsewhere [35,36].

Overall, some of the trials surveyed did not report information on important aspects of the study. Adherence to the CONSORT reporting guidelines [37] will improve communicating of what was done, allowing a better interpretation of the results.

As can be seen in Table 5, there were differences between trials in terms of the population included, the analytical approach, the quantity of missing data at the end of the study or the type of adverse events. This is not unexpected, since the design and analysis of the trials are adapted to the nature of the intervention. However, key aspects in the interpretation of any study include the need to assess the application of the results to the particular patient and the judicious evaluation of both efficacy and safety.

Table 5. Summary of some of the topics discussed.

	GATHER1	DERBY/OAKS	CHROMA/SPECTRI	SEATTLE	GATE
Metric of PE	mm/year	mm^2/year	mm^2/year	mm^2/year	mm^2/year
Adjusted PE	No	Yes	Yes	Yes	No
Atrophy location	Extrafoveal	Any	Any	Any	Any
Intention to treat	Yes	Yes	Yes	Yes	No?
Discontinuation, %	20.1	11.4	7.9	37.0	29.9
Specific adverse events	eAMD	eAMD	Increased IOP	Delayed DA	Reduced VA

DA: dark adaptation; eAMD: exudative age-related macular degeneration; IOP: intraocular pressure; PE: primary endpoint; VA: best-corrected visual acuity.

Readers should also bear in mind that small studies have less precision and wider confidence intervals (Figure 3), and thus can only report statistically significant results if the effect size is very large. When combined with publication bias (the trend of publishing studies with positive results), published small trials tend to overestimate the effects of therapy [38]. This overestimation of effect occurred, for example, in the lampalizumab and pegcetacoplan trials.

Geographic atrophy is a notable exception in research in macular diseases, inasmuch as the primary endpoint in all phase 3 clinical trials conducted thus far has been based on an anatomic change, which is a surrogate endpoint. In the case of the DERBY/OAKS trials, the positive effect of pegcetacoplan on the growth of area of atrophy was observed a few months after starting treatment, while its beneficial effect on visual function may take a few years. In addition, this therapy can increase the risk of short-term visual acuity loss due to incident eAMD. This is a key issue, considering that patients with this condition are elderly subjects with a relatively short life expectancy, estimated as 6.4 years in one study [39]. Adherence to therapy will also play an important role and will be influenced (among other things) by the cost of therapy, which may be close to 2000 USD/injection. All things considered, the benefits of therapy from the patient's perspective are, at least, debatable. In this context, PROMs may provide critical information, while economic evaluation through cost-utility (estimation of cost-per-QALY (quality-adjusted life year, one year lived in perfect health)) and cost-effectiveness analysis (comparison of costs and outcomes with and without treatment) will provide a useful perspective of the effects of this intervention for all those involved.

In summary, we reviewed some methodological aspects of phase 3 clinical trials. When applied to GA, similarities and differences between studies were noted. Poorly reported topics were highlighted, paving the way for improvement in future studies.

Author Contributions: Conceptualization, M.B.; methodology, M.B.; software, M.B.; validation, M.B., X.G.-S., A.G.-B., M.G., G.L., E.L., S.R., M.V. and L.S.; formal analysis, M.B.; investigation, M.B., X.G.-S., A.G.-B., M.G., G.L., E.L., S.R., M.V. and L.S.; data curation, M.B; writing—original draft preparation, M.B.; writing—review and editing, M.B., X.G.-S., A.G.-B., M.G., G.L., E.L., S.R., M.V. and L.S.; visualization, M.B., X.G.-S., A.G.-B., M.G., G.L., E.L., S.R., M.V. and L.S.; supervision, M.B.; project administration, M.B. All authors have read and agreed to the published version of the manuscript.

Funding: This research received no external funding.

Institutional Review Board Statement: Not applicable.

Informed Consent Statement: Not applicable.

Data Availability Statement: Not applicable.

Conflicts of Interest: The authors declare no conflict of interest.

References

1. Bhutto, I.; Lutty, G. Understanding age-related macular degeneration (AMD): Relationships between the photoreceptor/retinal pigment epithelium/Bruch's membrane/choriocapillaris complex. *Mol. Asp. Med.* **2012**, *33*, 295–317. [CrossRef] [PubMed]
2. Sayegh, R.G.; Kiss, C.G.; Simader, C.; Kroisamer, J.; Montuoro, A.; Mittermüller, T.J.; Azhary, M.; Bolz, M.; Kreil, D.P.; Schmidt-Erfurth, U. A systematic correlation of morphology and function using spectral domain optical coherence tomography and microperimetry in patients with geographic atrophy. *Br. J. Ophthalmol.* **2014**, *98*, 1050–1055. [CrossRef] [PubMed]
3. Sunness, J.S.; Gonzalez-Baron, J.; Applegate, C.A.; Bressler, N.M.; Tian, Y.; Hawkins, B.; Barron, Y.; Bergman, A. Enlargement of atrophy and visual acuity loss in the geographic atrophy form of age-related macular degeneration. *Ophthalmology* **1999**, *106*, 1768–1779. [CrossRef] [PubMed]
4. Meleth, A.D.; Mettu, P.; Agrón, E.; Chew, E.Y.; Sadda, S.R.; Ferris, F.L.; Wong, W.T. Changes in retinal sensitivity in geographic atrophy progression as measured by microperimetry. *Investig. Ophthalmol. Vis. Sci.* **2011**, *52*, 1119–1126. [CrossRef]
5. Chakravarthy, U.; Bailey, C.C.; Johnston, R.L.; McKibbin, M.; Khan, R.S.; Mahmood, S.; Downey, L.; Dhingra, N.; Brand, C.; Brittain, C.J.; et al. Characterizing Disease Burden and Progression of Geographic Atrophy Secondary to Age-Related Macular Degeneration. *Ophthalmology* **2018**, *125*, 842–849. [CrossRef]
6. Jaffe, G.J.; Westby, K.; Csaky, K.G.; Monés, J.; Pearlman, J.A.; Patel, S.S.; Joondeph, B.C.; Randolph, J.; Masonson, H.; Rezaei, K.A. C5 Inhibitor Avacincaptad Pegol for Geographic Atrophy Due to Age-Related Macular Degeneration: A Randomized Pivotal Phase 2/3 Trial. *Ophthalmology* **2021**, *128*, 576–586. [CrossRef]
7. Apellis Website. Available online: https://investors.apellis.com/events-and-presentations (accessed on 20 March 2023).
8. Holz, F.G.; Sadda, S.R.; Busbee, B.; Chew, E.Y.; Mitchell, P.; Tufail, A.; Brittain, C.; Ferrara, D.; Gray, S.; Honigberg, L.; et al. Efficacy and Safety of Lampalizumab for Geographic Atrophy Due to Age-Related Macular Degeneration. *JAMA Ophthalmol.* **2018**, *136*, 666. [CrossRef]
9. Rosenfeld, P.J.; Dugel, P.U.; Holz, F.G.; Heier, J.S.; Pearlman, J.A.; Novack, R.L.; Csaky, K.G.; Koester, J.M.; Gregory, J.K.; Kubota, R. Emixustat Hydrochloride for Geographic Atrophy Secondary to Age-Related Macular Degeneration: A Randomized Clinical Trial. *Ophthalmology* **2018**, *125*, 1556–1567. [CrossRef]
10. Jaffe, G.J.; Schmitz-Valckenberg, S.; Boyer, D.; Heier, J.; Wolf-Schnurrbusch, U.; Staurenghi, G.; Schmidt-Erfurth, U.; Holz, F.G. Randomized Trial to Evaluate Tandospirone in Geographic Atrophy Secondary to Age-Related Macular Degeneration: The GATE Study. *Am. J. Ophthalmol.* **2015**, *160*, 1226–1234. [CrossRef]
11. Sadda, S.R.; Guymer, R.; Holz, F.G.; Schmitz-Valckenberg, S.; Curcio, C.A.; Bird, A.C.; Blodi, B.A.; Bottoni, F.; Chakravarthy, U.; Chew, E.Y.; et al. Consensus Definition for Atrophy Associated with Age-Related Macular Degeneration on OCT: Classification of Atrophy Report 3. *Ophthalmology* **2018**, *125*, 537–548. [CrossRef]
12. Villani, E.; Vujosevic, S. Foreword: Biomarkers and surrogate endpoints in ophthalmic clinical research. *Investig. Ophthalmol. Vis. Sci.* **2017**, *58*, BIOi–BIOii. [CrossRef]
13. Csaky, K.; Ferris, F.; Chew, E.Y.; Nair, P.; Cheetham, J.K.; Duncan, J.L. Report From the NEI/FDA Endpoints Workshop on Age-Related Macular Degeneration and Inherited Retinal Diseases. *Investig. Opthalmol. Vis. Sci.* **2017**, *58*, 3456. [CrossRef]
14. Yehoshua, Z.; Rosenfeld, P.J.; Gregori, G.; Feuer, W.J.; Falcão, M.; Lujan, B.J.; Puliafito, C. Progression of geographic atrophy in age-related macular degeneration imaged with spectral domain optical coherence tomography. *Ophthalmology* **2011**, *118*, 679–686. [CrossRef]
15. Schmetterer, L.; Scholl, H.; Garhöfer, G.; Janeschitz-Kriegl, L.; Corvi, F.; Sadda, S.R.; Medeiros, F.A. Endpoints for clinical trials in ophthalmology. *Prog. Retin. Eye Res.* 2023; *in press*. [CrossRef]
16. Gregson, J.; Sharples, L.; Stone, G.W.; Burman, C.-F.; Öhrn, F.; Pocock, S. Nonproportional Hazards for Time-to-Event Outcomes in Clinical Trials: JACC Review Topic of the Week. *J. Am. Coll. Cardiol.* **2019**, *74*, 2102–2112. [CrossRef]

17. Kersten, E.; Geerlings, M.J.; Pauper, M.; Corominas, J.; Bakker, B.; Altay, L.; Fauser, S.; De Jong, E.K.; Hoyng, C.B.; Hollander, A.I.D. Genetic screening for macular dystrophies in patients clinically diagnosed with dry age-related macular degeneration. *Clin. Genet.* **2018**, *94*, 569–574. [CrossRef]
18. Fleckenstein, M.; Mitchell, P.; Freund, K.B.; Sadda, S.; Holz, F.G.; Brittain, C.; Henry, E.C.; Ferrara, D. The Progression of Geographic Atrophy Secondary to Age-Related Macular Degeneration. *Ophthalmology* **2018**, *125*, 369–390. [CrossRef]
19. Biarnés, M.; Arias, L.; Alonso, J.; Garcia, M.; Hijano, M.; Rodríguez, A.; Serrano, A.; Badal, J.; Muhtaseb, H.; Verdaguer, P.; et al. Increased Fundus Autofluorescence and Progression of Geographic Atrophy Secondary to Age-Related Macular Degeneration: The GAIN Study. *Am. J. Ophthalmol.* **2015**, *160*, 345–353.e5. [CrossRef]
20. Biarnés, M.; Colijn, J.M.; Sousa, J.; Ferraro, L.L.; Garcia, M.; Verzijden, T.; Meester-Smoor, M.A.; Delcourt, C.; Klaver, C.C.W.; den Hollander, A.I.; et al. Genotype- and Phenotype-Based Subgroups in Geographic Atrophy Secondary to Age-Related Macular Degeneration: The EYE-RISK Consortium. *Ophthalmol. Retin.* **2020**, *4*, 1129–1137. [CrossRef]
21. Wei, W.; Mazzola, M.; Otero-Marquez, O.; Tong, Y.; Souied, E.; Querques, G.; Freund, K.B.; Smith, R.T. Two potentially distinct pathways to geographic atrophy in age-related macular degeneration characterized by quantitative fundus autofluorescence. *Eye* 2023, *in press*. [CrossRef]
22. Yaspan, B.L.; Williams, D.F.; Holz, F.G.; Regillo, C.D.; Li, Z.; Dressen, A.; Campagne, M.V.L.; Le, K.N.; Graham, R.R.; Beres, T.; et al. Targeting factor D of the alternative complement pathway reduces geographic atrophy progression secondary to age-related macular degeneration. *Sci. Transl. Med.* **2017**, *9*, eaaf1443. [CrossRef] [PubMed]
23. Yehoshua, Z.; De Amorim Garcia Filho, C.A.; Nunes, R.P.; Gregori, G.; Penha, F.M.; Moshfeghi, A.A.; Luo, H.; Kang, Z.; Sadda, S.; Feuer, W.; et al. Association between Growth of Geographic Atrophy and the Complement Factor i Locus. *Ophthalmic Surg Lasers Imaging Retin.* **2015**, *46*, 772–774. [CrossRef] [PubMed]
24. Lachin, J.M. Fallacies of last observation carried forward analyses. *Clin. Trials* **2016**, *13*, 161–168. [CrossRef] [PubMed]
25. Steinle, N.C.; Pearce, I.; Monés, J.; Metlapally, R.; Saroj, N.; Hamdani, M.; Ribeiro, R.; Rosenfeld, P.J.; Lad, E.M. Impact of Baseline Characteristics on Geographic Atrophy Progression in the FILLY Trial Evaluating the Complement C3 Inhibitor Pegcetacoplan. *Am. J. Ophthalmol.* **2021**, *227*, 116–124. [CrossRef] [PubMed]
26. Liao, D.S.; Grossi, F.V.; El Mehdi, D.; Gerber, M.R.; Brown, D.M.; Heier, J.S.; Wykoff, C.C.; Singerman, L.J.; Abraham, P.; Grassmann, F.; et al. Complement C3 Inhibitor Pegcetacoplan for Geographic Atrophy Secondary to Age-Related Macular Degeneration: A Randomized Phase 2 Trial. *Ophthalmology* **2020**, *127*, 186–195. [CrossRef]
27. Faraone, S.V. Interpreting estimates of treatment effects: Implications for managed care. *P T* **2008**, *33*, 700–711.
28. Smith, V.A.; Coffman, C.J.; Hudgens, M.G. Interpreting the Results of Intention-to-Treat, Per-Protocol, and As-Treated Analyses of Clinical Trials. *JAMA J. Am. Med. Assoc.* **2021**, *326*, 433–434. [CrossRef]
29. Ioannidis, J.P.; Lau, J. Pooling research results: Benefits and limitations of meta-analysis. *Jt. Comm. J. Qual. Improv.* **1999**, *25*, 462–469. [CrossRef]
30. Pocock, S.J.; Collier, T.J. Statistical Appraisal of 6 Recent Clinical Trials in Cardiology: JACC State-of-the-Art Review. *J. Am. Coll. Cardiol.* **2019**, *73*, 2740–2755. [CrossRef]
31. Monés, J.; Srivastava, S.K.; Jaffe, G.J.; Tadayoni, R.; Albini, T.A.; Kaiser, P.K.; Holz, F.G.; Korobelnik, J.-F.; Kim, I.K.; Pruente, C.; et al. Risk of Inflammation, Retinal Vasculitis, and Retinal Occlusion–Related Events with Brolucizumab: Post Hoc Review of HAWK and HARRIER. *Ophthalmology* **2021**, *128*, 1050–1059. [CrossRef]
32. Reibaldi, M.; Pulvirenti, A.; Avitabile, T.; Bonfiglio, V.; Russo, A.; Mariotti, C.; Bucolo, C.; Mastropasqua, R.; Parisi, G.; Longo, A. Pooled estimates of incidence of endophthalmitis after intravitreal injection of anti–vascular endothelial growth factor agents with and without topical antibiotic prophylaxis. *Retina* **2018**, *38*, 1–11. [CrossRef]
33. Saksens, N.T.; Fleckenstein, M.; Schmitz-Valckenberg, S.; Holz, F.G.; den Hollander, A.I.; Keunen, J.E.; Boon, C.J.; Hoyng, C.B. Macular dystrophies mimicking age-related macular degeneration. *Prog. Retin. Eye Res.* **2014**, *39*, 23–57. [CrossRef]
34. Holz, F.G.; Tadayoni, R.; Beatty, S.; Berger, A.; Cereda, M.G.; Cortez, R.; Hoyng, C.B.; Hykin, P.; Staurenghi, G.; Heldner, S.; et al. Multi-country real-life experience of anti-vascular endothelial growth factor therapy for wet age-related macular degeneration. *Br. J. Ophthalmol.* **2015**, *99*, 220–226. [CrossRef]
35. Pocock, S.J.; Stone, G.W. The Primary Outcome Fails—What Next? *N. Engl. J. Med.* **2016**, *375*, 861–870. [CrossRef]
36. Pocock, S.J.; Stone, G.W. The Primary Outcome Is Positive—Is That Good Enough? *N. Engl. J. Med.* **2016**, *375*, 971–979. [CrossRef]
37. Moher, D.; Hopewell, S.; Schulz, K.F.; Montori, V.; Gøtzsche, P.C.; Devereaux, P.J.; Elbourne, D.; Egger, M.; Altman, D.G. CONSORT 2010 explanation and elaboration: Updated guidelines for reporting parallel group randomised trials. *Int. J. Surg.* **2012**, *10*, 28–55. [CrossRef]
38. Ioannidis, J.P.A.; Pereira, T.V.; Horwitz, R.I. Emergence of large treatment effects from small trials—Reply. *JAMA* **2013**, *309*, 768–769. [CrossRef]
39. Colijn, J.M.; Liefers, B.; Joachim, N.; Verzijden, T.; Meester-Smoor, M.A.; Biarnés, M.; Monés, J.; de Jong, P.T.V.M.; Vingerling, J.R.; Mitchell, P.; et al. Enlargement of Geographic Atrophy from First Diagnosis to End of Life. *JAMA Ophthalmol.* **2021**, *139*, 743. [CrossRef]

biomedicines

MDPI

Article

Quantitative Autofluorescence in Non-Neovascular Age Related Macular Degeneration

Shruti Chandra [1,2], Manjot K. Grewal [2], Sarega Gurudas [1], Rajan Sondh [3], Alan Bird [1,2], Glen Jeffery [1], Victor Chong [1] and Sobha Sivaprasad [1,2,*]

[1] Institute of Ophthalmology, University College London, London EC1V 9EL, UK
[2] Moorfields National Institute of Health and Care Research Facility, Moorfields Eye Hospital, London EC1V 2PD, UK
[3] Sandwell and West Birmingham Hospitals NHS Trust, Birmingham B71 4HJ, UK
* Correspondence: sobha.sivaprasad@nhs.net

Abstract: Quantitative autofluorescence (qAF8) level is a presumed surrogate marker of lipofuscin content in the retina. We investigated the changes in the qAF8 levels in non-neovascular AMD. In this prospective cohort study, Caucasians aged ≥50 years with varying severity of non-neovascular AMD in at least one eye and Snellen visual acuity ≥6/18 were recruited. The qAF8 levels were analysed in the middle eight segments of the Delori pattern (HEYEX software, Heidelberg, Germany). The AMD categories were graded using both the Beckman classification and multimodal imaging (MMI) to include the presence of subretinal drusenoid deposits (SDD). A total of 353 eyes from 231 participants were analyzed. Compared with the age-matched controls, the qAF8 values decreased in the eyes with AMD (adjusted % difference = −19.7% [95% CI −28.8%, −10.4%]; $p < 0.001$) and across the AMD categories, (adjusted % differences; Early, −13.1% (−24.4%, −1%), $p = 0.04$; intermediate AMD (iAMD), −22.9% (−32.3%, −13.1%), $p < 0.001$; geographic atrophy −25.2% (−38.1%, −10.4%), $p = 0.002$). On MMI, the qAF8 was reduced in the AMD subgroups relative to the controls, (adjusted % differences; Early, −5.8% (−18.9%, 8.3%); $p = 0.40$; iAMD, −26.7% (−36.2%, −15.6%); $p < 0.001$; SDD, −23.7% (−33.6%, −12.2%); $p < 0.001$; atrophy, −26.7% (−39.3%, −11.3%), $p = 0.001$). The qAF8 levels declined early in AMD and were not significantly different between the severity levels of non-neovascular AMD, suggesting the early and sustained loss of function of the retinal pigment epithelium in AMD.

Keywords: age-related macular degeneration; quantitative autofluorescence; subretinal drusenoid deposits; hyperreflective foci; drusen; geographic atrophy

Citation: Chandra, S.; Grewal, M.K.; Gurudas, S.; Sondh, R.; Bird, A.; Jeffery, G.; Chong, V.; Sivaprasad, S. Quantitative Autofluorescence in Non-Neovascular Age Related Macular Degeneration. *Biomedicines* **2023**, *11*, 560. https://doi.org/10.3390/biomedicines11020560

Academic Editor: Oyuna Kozhevnikova

Received: 30 January 2023
Accepted: 14 February 2023
Published: 15 February 2023

1. Introduction

Bisretinoid fluorophores in the phagocytosed photoreceptor outer segments are by-products of the visual cycle that accumulate as lipofuscin within the human retinal pigment epithelium (RPE). Lipofuscin is a major contributor of fundus autofluorescence (FAF). An age-related non-linear increase in the amount of autofluorescence in RPE has been observed, with significant variations in the same age-group [1]. Delori et al. confirmed that lipofuscin fluorescence increases linearly until age 70 in healthy eyes, then declines, probably due to the loss of RPE cells [2].

Excessive lipofuscin could adversely affect the essential RPE functions and contribute to the pathogenesis of age-related macular degeneration (AMD) [3–5]. The phototoxic and proinflammatory effects of lipofuscin may cause RPE cell death in AMD eyes [6,7]. For example, hyper-autofluorescence, seen around the border of geographic atrophy (GA), may be a sign of excessive lipofuscin that precedes atrophy or a displacement or stacking up of RPE cells with the accumulated lipofuscin [8]. The presence of lipofuscin in drusen also suggests that it may be involved in drusen biogenesis [9]. By contrast, the loss of lipofuscin from RPE cell death is seen as hypo-autofluorescence, as is typically seen in GA.

Quantitative fundus autofluorescence (qAF8) enables the in vivo quantification of these FAF signals compared with an internal standard. It has been concluded that eyes with increased qAF8 levels may benefit from anti-lipofuscin therapies [5,10,11]. Studies have reported qAF8 levels in AMD eyes and its correlation with factors such as age, sex, ethnicity, smoking status, type and volume of drusen [12–15]. These reports consistently showed either normal or reduced qAF8 levels in AMD eyes compared with age-matched controls. It is unclear whether the reduced qAF8 levels are associated with the loss of function and integrity of RPE cells or because less lipofuscin is produced due to abnormalities in the visual cycle [16]. As subretinal drusenoid deposits (SDD) are associated with a prolonged rod-intercept time, it may be a surrogate for impaired visual cycle.

Deciphering the relation of qAF8 to the RPE and the visual cycle is important as there are interventional trials for GA aimed to modulate the visual cycle or reduce lipofuscin. Examples include oral LBS-008 (Tinlarebant), a small molecule retinol binding protein 4 (RBP4) antagonist that is being evaluated in Stargardt Disease. Its effect on AMD is unknown. Another RBP4 inhibitor, fenretinide, showed a correlation with a reduction in the GA growth rate, but did not delay the progression of GA [17]. Another phase three trial on ALK-001, a chemically modified vitamin A, is also being investigated for GA (NCT03845582). Furthermore, retinylamine, which lowers the concentration of all-trans retinal, has also been considered for AMD [18]. Another visual cycle modulator, Emixustat (NCT01802866), which inhibits RPE-specific protein RPE65 was shown to reduce the growth rate of small sized atrophy due to Stargardt disease but failed to show a benefit in GA [19]. The results of these studies show that further work is required to understand how anti-lipofuscin drugs will benefit eyes with AMD.

Further studies dissecting the qAF8 levels in ageing and various stages of AMD may provide new insight in AMD therapies. Using multimodal imaging, evaluating the differences of qAF8 levels in eyes with and without SDD may indirectly provide information on the relation of qAF8 and the visual cycle. In addition, the relation of qAF8 with the worsening health of RPE cells may be interrogated by its associations with optical coherent tomography (OCT) features, such as hyperreflective foci (HRF), incomplete and complete RPE and outer retinal atrophy (iRORA and cRORA, respectively) [20].

The aims of this study were to: (i) evaluate whether qAF8 levels vary with AMD progression; (ii) study whether qAF8 levels vary in eyes with and without SDD to investigate the relation of qAF8 with the visual cycle and (iii) interrogate whether OCT risk features of RPE cell death influence the qAF8 levels.

2. Materials and Methods

This single centre, prospective cohort study (PEONY study) was approved by the London-Chelsea Research Ethics Committee London REC 19/LO/0931. Written informed consent was obtained from all participants and the study followed the tenets of the Declaration of Helsinki.

2.1. Setting

The National Institute of Health and Research Clinical Research Facility at a tertiary care hospital in London, UK.

2.2. Participants

Caucasians aged ≥50 years with varying severity of non-neovascular AMD in at least one eye and age-matched controls with healthy maculae were recruited for this study between August 2020 and July 2022. The inclusion criteria for non-neovascular AMD in at least one eye were a Snellen visual acuity of ≥6/18 with media clarity for qAF8 and other imaging. The exclusion criteria included any condition that, in the opinion of the investigator, could affect or alter visual acuity or retinal imaging, such as vitreous opacities, epiretinal membrane and other comorbid conditions such as diabetic retinopathy.

2.3. Study Assessments

Age, gender and lens status were recorded. The multimodal imaging obtained for this study included near infrared reflectance (NIR), fundus autofluorescence (FAF), qAF8, en-face SD-OCT scans (97-line scans, 6 × 6 mm grid), enhanced depth imaging OCT (EDI-OCT), OCT angiography. These were conducted on Spectralis HRA + OCT, Heidelberg Engineering, Germany.

2.4. Image Acquisition

Two trained operators acquired the images after mydriasis. A standard operating procedure was followed. In summary, with room lights were turned off and the NIR image was recorded first. In the qAF8 mode, the image was refocused to ensure a uniform signal over the entire field. During the set-up and focusing in the qAF8 mode, the fundus was exposed for 20 s to reduce visual pigment absorption. Each image was acquired in high-speed video mode, at least 2 images were recorded, each being 12 frames. All video images were examined for image quality and at least 9 of the 12 frames were selected. The selected frames were aligned and the mean image saved. The image quality was graded based on the criteria defined by von der Emde et al. (poor, acceptable or excellent) and all images with quality acceptable or above were included [21].

2.5. Image Analysis

The qAF8 images were analysed using the HEYEX software. The inter-operator agreement of qAF8 on five participants was measured first (kappa-0.89). The qAF8 levels were measured from the qAF8 segments, which refer to the middle ring of the Delori grid centered on the fovea that divides the area between 9° to 11° eccentricity from the fovea into eight segments. Vessels were automatically excluded from the analysis by the software. The threshold setting was manually adjusted if necessary. Phakic eyes were subsequently corrected for normative age-related optical media density [13,22]. No age-adjustment for ocular media absorption was applied for pseudophakic eyes [22].

The normalised grey values from the middle eight segments of the qAF8 image were averaged to provide the main outcome measure of this study (qAF8).

2.6. Definition of Various AMD Categories

The qAF8 levels were compared across the various stages of AMD on colour photographs based on the Beckman classification [23]. The stages included: (i) no evidence of AMD or normal aging was defined as the presence of small drusen or druplets of <63 μm diameter; (ii) early AMD included eyes with a medium drusen, between 63 μm and 124 μm in diameter, with no pigmentary changes; (iii) intermediate AMD (iAMD) was defined as eyes with a drusen diameter of ≥125 μm or medium drusen with pigmentary changes; and (iv) eyes with GA [23]. This allowed us to investigate whether the qAF8 levels varied with a visible progressive loss of integrity of RPE on the colour photographs.

Next, qAF8 was compared in the AMD categories classified by the multimodal imaging (MMI). Participants with a minimum of 5 SDD on different line scans on OCT confirmed on NIR were identified as a separate category, irrespective of the size of the drusen present. Eyes with SDD were further divided into stages: stage 1, defined as a diffuse deposition of granular hyperreflective material between the RPE-Bruch membrane band and the ellipsoid zone (EZ); stage 2, mounds of accumulated material sufficient to deflect inwardly the contour of the EZ; and stage 3, the presence of a conical appearance and breaking through the EZ [24]. AMD eyes without SDD were grouped based on their drusen size on the OCT, as reported by Kim et al. [25,26]. On the en-face OCT, a large drusen was defined as a drusen with a diameter ≥145 μm; the medium drusen diameters were between 100 μm and 144 μm; and the small drusen had diameters <100 μm on the OCT. Accordingly, the multimodal imaging classification of AMD included: (i) early AMD on the OCT, defined as maximum druse diameter <100 μm with no SDD or atrophy; (ii) intermediate AMD defined as having more than 1 druse with diameter 100–145 μm or at least 1 druse (>145 μm) without SDD or atrophy (cRORA), but may have HRF or iRORA; (iii) eyes with SDD; and (iv) eyes with atrophy (cRORA). We also investigated the relation of the drusen volume with the qAF8 values. The OCT scans were segmented by in-built automated Heidelberg software and manually corrected where necessary.

Lastly, we compared the qAF8 values in the AMD eyes with varying degress of RPE cell loss identified on the OCT. These included comparing the qAF8 in eyes with: (i) no HRF, iRORA or cRORA; (ii) only HRF with no iRORA or cRORA; (iii) presence of iRORA with no cRORA; and (iv) cRORA.

2.7. Statistical Analysis

The data were summarised using mean (SD) or median (IQR) for continuous variables and number (%) for categorical variables. The statistical differences in the $\log_e$ Mean qAF8 levels between the controls and individual AMD groups by multimodal imaging OCT and Beckman classification were compared after adjustment for the potential confounders of age, sex and lens status (pseudophakic vs. phakic). The stage of SDD, qAF8 and drusen volume were summarised based on the varying severity of RPE cell loss in AMD. The associations after adjusting for age, sex and lens status were quantified with odds ratio (95% CI) and *p*-value. Generalised Estimating Equations (GEE) with an exchangeable working correlation structure were used to account for the within-participant correlation among those with data from both eyes [27,28]. The qAF8 levels were $\log_e$-transformed to avoid estimating misleading associations from outliers. Model coefficient estimates were therefore interpreted in relation to the relative percentage increase rather than the absolute increase in the qAF8 values. Where log-transformations were applied to the independent variable qAF8 in the GEE models, the model estimates were interpreted in relation to the relative percentage increase in the independent variable. Spearmans correlation coefficient for clustered data, using the within-cluster resampling (WCR)-based method and standard errors based on fixed denominator, was used to test the correlation between age and mean qAF8 levels [29].

3. Results

3.1. Participant and Imaging Characteristics

The analysis sample included 105 eyes from 63 controls and 248 eyes from 176 AMD participants (Figure S1). These participants had a mean age of 70.3 years (SD 8.5), 138 were females (59.7%) and 60 (17.0%) eyes were pseudophakic (Table 1).

Table 1. Participants and imaging characteristics in AMD categories and age-matched controls.

		AMD vs. Controls	
Variable	**Overall, N = 353 Eyes of 231 Participants**	**Controls, N = 105 Eyes of 63 Participants**	**AMD, N = 248 Eyes of 176 Participants**
Participant level [a]			
Number of participants with bilateral eyes	122 (52.8%)	42 (66.7%)	72 (40.9%)
Age, years	70.3 (8.5)	64.2 (8.2)	72.4 (7.5)
Age, years			
<60	30 (13.0%)	20 (31.8%)	11 (6.3%)
60–69	65 (28.1%)	24 (38.1%)	44 (25.0%)
70–79	104 (45.0%)	15 (23.8%)	92 (52.3%)
≥80	32 (13.9%)	4 (6.3%)	29 (16.5%)
Gender			
M	93 (40.3%)	28 (44.4%)	68 (38.6%)
F	138 (59.7%)	35 (55.6%)	108 (61.4%)

Table 1. *Cont.*

Variable	Overall, N = 353 Eyes of 231 Participants	AMD vs. Controls	
		Controls, N = 105 Eyes of 63 Participants	**AMD, N = 248 Eyes of 176 Participants**
Eye level [b]			
Pseudophakic eyes	60 (17.0%)	7 (6.7%)	53 (21.4%)
Beckman AMD classification			
Normal	105 (29.7%)	105 (100.0%)	0 (0.0%)
Early	59 (16.7%)	0 (0.0%)	59 (23.8%)
iAMD	161 (45.6%)	0 (0.0%)	161 (64.9%)
GA	28 (7.9%)	0 (0.0%)	28 (11.3%)
AMD classification based on multimodal imaging			
Normal	105 (29.7%)	105 (100.0%)	0 (0.0%)
Early	47 (13.3%)	0 (0.0%)	47 (19.0%)
iAMD	76 (21.5%)	0 (0.0%)	76 (30.6%)
SDD	97 (27.5%)	0 (0.0%)	97 (39.1%)
cRORA	28 (7.9%)	0 (0.0%)	28 (11.3%)
Median qAF8 (IQR)	184.0 (132.2, 246.4)	241.6 (185.4, 294.4)	163.5 (122.6, 216.6)
Analysis in AMD eyes [c]; n (%)			
Stage of SDD			
0	151/248 (60.9%)	0 (NA%)	151/248 (60.9%)
1	5/248 (2.0%)	0 (NA%)	5/248 (2.0%)
2	15/248 (6.0%)	0 (NA%)	15/248 (6.0%)
3	77/248 (31.0%)	0 (NA%)	77/248 (31.0%)
Indicators of RPE integrity			
Absence of HRF, iRORA or cRORA	153/248 (61.7%)	0 (NA%)	153/248 (61.7%)
Presence of HRF without evidence of iRORA or cRORA	16/248 (6.5%)	0 (NA%)	16/248 (6.5%)
Presence of iRORA without any evidence of cRORA	51/248 (20.6%)	0 (NA%)	51/248 (20.6%)
Presence of cRORA	28/248 (11.3%)	0 (NA%)	28/248 (11.3%)
Drusen volume in those without cRORA (N = 220 eyes)	0.44 (0.36, 0.54)	NA	0.44 (0.36, 0.54)

[a] Participant level characteristics were based on total cohort of 231 participants. The number of participants across healthy and AMD do not add up to 231 (total number of participants in the cohort) due to bilateral eligibility in the cohort where eyes from the same participant may belong in both groups. [b] Eye level characteristics based on total analysed sample of 353 eyes. [c] Analysis in AMD eyes based on 248 eyes with AMD only, except drusen volume which has been analysed in 220 AMD eyes without cRORA. Abbreviations: AMD—age related macular degeneration; cRORA—complete retinal pigment epithelium and outer retinal atrophy; GA—Geographic atrophy; HRF—Hyperreflective foci; iAMD—intermediate AMD; iRORA—incomplete retinal pigment epithelium and outer retinal atrophy; IQR—Interquartile range; qAF8—quantitative autofluorescence; RPE—Retinal pigment epithelium; SDD—subretinal drusenoid deposits.

3.2. Comparison of qAF8 Levels in AMD versus Control Eyes Aged ≥50 Years

The spearman correlation coefficient between age and qAF8 values in controls ≥50 years was $\rho_{WCR} = -0.21$; $p = 0.08$, in AMD eyes ≥50 years was $\rho_{WCR} = -0.05$; $p = 0.49$ (Figure S2) and in ≥65 years was $\rho_{WCR} = -0.05$; $p = 0.54$. Compared with the controls ≥50 years, the qAF8 values were significantly reduced in eyes with any AMD (241.6 [IQR 185.4–294.4] vs. 163.5 [IQR 122.6–216.6]; unadjusted difference= −25.2% [95% CI −33.0%, −16.5%]; $p < 0.001$; GEE model). The participant and imaging characteristics in AMD categories are presented in Table S1.

3.3. Comparison of qAF8 Levels in AMD Categories Based on Beckman Classification versus Control Eyes from Participants Aged ≥50 Years

There was a significant reduction in qAF8 in the AMD eyes versus control (−19.7% [95% CI −28.8%, −10.4%]; $p < 0.001$, Table 2). Statistical differences were found among all of the Beckman subgroups relative to the control eyes, in both the univariate analysis and following the adjustment for age, gender and lens status (adjusted % differences; Early AMD, −13.1% (−24.4%, −1%), $p = 0.04$; iAMD, −22.9% (−32.3%, −13.1%), $p < 0.001$; and GA −25.2% (−38.1%, −10.4%), $p = 0.002$). Significant differences were identified comparing early AMD with iAMD (% adjusted difference −11.4% [95% CI −0.10%, −21.3%]; $p = 0.048$, while the comparison between GA and early AMD failed to reach statistical significance in both the univariate and adjusted analysis (% adjusted difference −14.0% [95% CI −28.7%, 3.6%]; $p = 0.11$). No statistically significant differences were found between iAMD and GA.

Table 2. Parameters associated with log-qAF8 mean adjusted for age, gender and lens status using GEE models.

Characteristics	N	Coefficient in Log-qAF8 Units	qAF8 Difference in %	*p*-Value
Controls vs. AMD	353			
Controls		-	-	
AMD		−0.22 (−0.34–−0.11)	−19.7% (−28.8%, −10.4%)	<0.001
Beckman classification	353			
Normal		-	-	
Early		−0.14 (−0.28–−0.01)	−13.1% (−24.4%, −1%)	0.04
iAMD		−0.26 (−0.39–−0.14)	−22.9% (−32.3%, −13.1%)	<0.001
GA		−0.29 (−0.48–−0.11)	−25.2% (−38.1%, −10.4%)	0.002
AMD classification based on multimodal imaging	353			
Normal		-		
Early without SDD		−0.06 (−0.21–0.08)	−5.8% (−18.9%, 8.3%)	0.40
iAMD without SDD		−0.31 (−0.45–−0.17)	−26.7% (−36.2%, −15.6%)	<0.001
SDD		−0.27 (−0.41–−0.13)	−23.7% (−33.6%, −12.2%)	<0.001
cRORA		−0.31 (−0.50–−0.12)	−26.7% (−39.3%, −11.3%)	0.001
Stage of SDD in those with SDD	97			
Stage 1 or 2		-		
Stage 3		−0.05 (−0.28–0.18)	−4.9% (−24.4%, 19.7%)	0.67
Stage of SDD	248			
0		-		
1		0.01 (−0.17–0.19)	1% (−15.6%, 20.9%)	0.88
2		0.05 (−0.22–0.33)	5.1% (−19.7%, 39.1%)	0.71
3		−0.05 (−0.17–0.07)	−4.9% (−15.6%, 7.3%)	0.38
Drusen volume in those without cRORA, per 0.1-unit increase	220	−0.01 (−0.05–0.02)	−1.4% (−4.6%, 1.9%)	0.40
Indicators of RPE integrity	248			
Absence of HRF without evidence of iRORA or cRORA		-		

Table 2. *Cont.*

Characteristics	N	Coefficient in Log-qAF8 Units	qAF8 Difference in %	p-Value
Presence of HRF without evidence of iRORA or cRORA		−0.18 (−0.43–0.07)	−16.5% (−34.9%, 7.25%)	0.16
Presence of iRORA without any evidence of cRORA		−0.16 (−0.28–−0.03)	−14.8% (−24.4%, −3.0%)	0.01
Presence of cRORA		−0.13 (−0.30–0.05)	−12.2% (−25.9%, 5.1%)	0.15

Abbreviations: AMD—age related macular degeneration; cRORA—complete retinal pigment epithelium and outer retinal atrophy; GA—Geographic atrophy; GEE—Generalised estimating Equation; HRF—Hyperreflective foci; iAMD—intermediate AMD; iRORA—incomplete retinal pigment epithelium and outer retinal atrophy; qAF8—quantitative autofluorescence; RPE—Retinal pigment epithelium; SDD—subretinal drusenoid deposits.

3.4. Comparison of qAF8 Levels in AMD Categories Based on MMI versus Control Eyes from Participants Aged ≥50 Years

The qAF8 levels were found to be reduced in the eyes with AMD compared to the normal eyes, while the iAMD, SDD and GA groups are in close correspondence (Figure 1). In the adjusted analysis, differences were found between the OCT AMD subgroups relative to the controls, with the exception of early AMD (adjusted % differences; Early AMD, −5.8% (−18.9%, 8.3%); $p = 0.40$; iAMD, −26.7% (−36.2%, −15.6%); $p < 0.001$; SDD, −23.7% (−33.6%, −12.2%); $p < 0.001$; and cRORA, −26.7% (−39.3%, −11.3%), $p = 0.001$, Table 2). Significant differences were found when comparing early AMD to the other categories of AMD based on the MMI classification (iAMD; % adjusted difference of −21.9% [95% CI −32.6%, −9.5%]; $p = 0.001$), SDD; −19.0% [95% CI −30.4%, −5.7%]; $p = 0.006$), cRORA −22.0% [95% CI −36.1%, −4.8%]; $p = 0.01$). No significant differences were found when comparing the non-early categories of AMD. Figure 2 shows example qAF8 images of the participants with the same age across the different AMD groups.

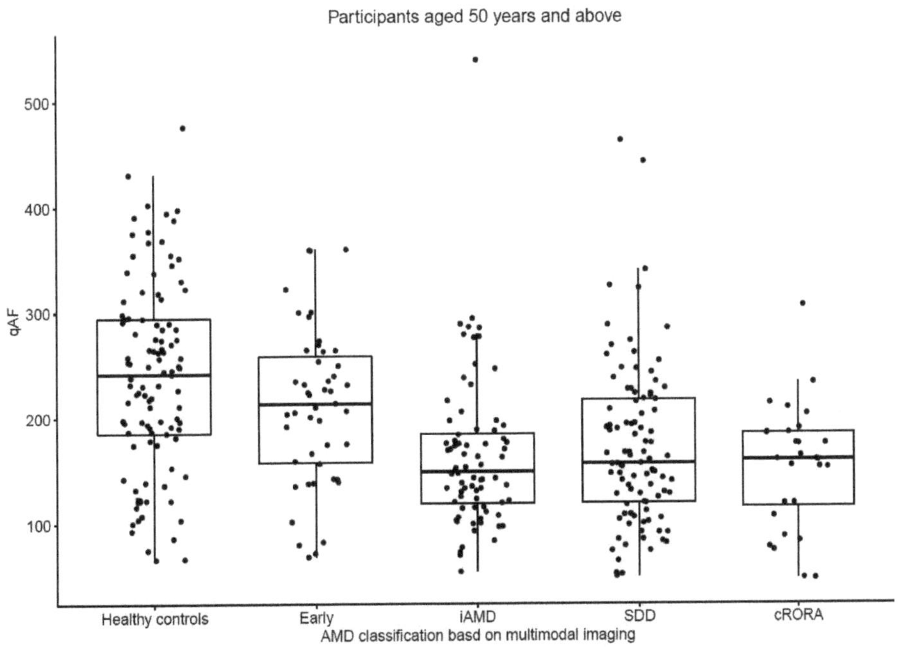

Figure 1. Mean qAF8 values in multimodal imaging categories of AMD versus age-matched controls. The multimodal imaging AMD classification shown on the x-axis. Comparison was with age-matched

controls. Early-drusen diameter <100 μm and no SDD, intermediate—drusen diameter 100–144 μm or 1 drusen 145 μm or more and no SDD, SDD-eyes with SDD with no cRORA and final group-cRORA. Abbreviations: AMD—age related macular degeneration; cRORA—complete retinal pigment epithelium and outer retinal atrophy; iAMD—intermediate AMD; qAF8—quantitative autofluorescence; SDD—subretinal drusenoid deposits.

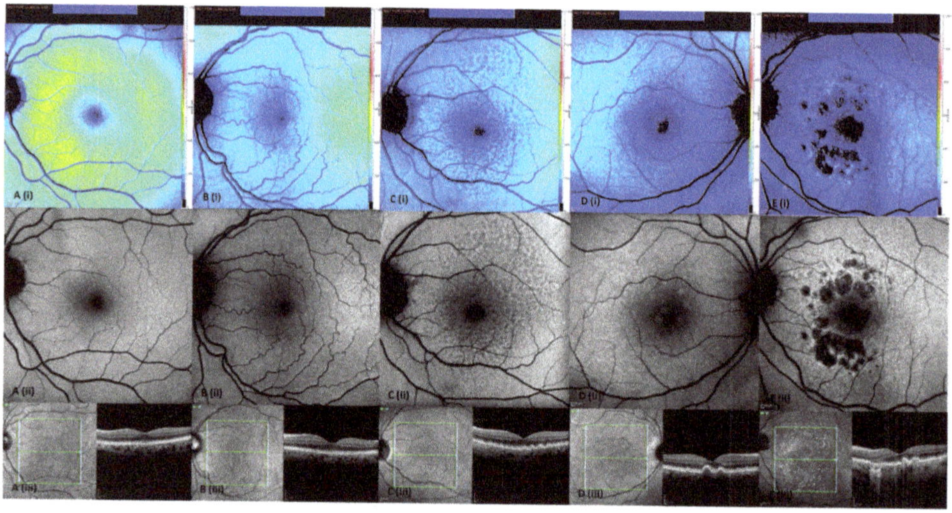

Figure 2. Multimodal imaging showing qAF8 changes across degeneration (AMD) categories in participants with same age. (**A**)—Controls; (**B**)—Early AMD; (**C**)—Subretinal drusenoid deposit; (**D**)—Intermediate AMD; (**E**)—Multifocal geographic atrophy. Images (i), (ii) and (iii) show the qAF8 image, fundus autofluorescence and IR + OCT images, respectively, for all groups. Abbreviations: AMD—age related macular degeneration; cRORA—complete retinal pigment epithelium and outer retinal atrophy; GA—Geographic atrophy; iAMD—intermediate AMD; IR- Infrared reflectance; OCT—Optical Coherence Tomography; qAF8—quantitative autofluorescence; SDD—subretinal drusenoid deposits.

3.5. Comparison of qAF8 Levels in AMD Eyes with and without Risk Factors of Progression to Geographic Atrophy or cRORA

Table S2 shows that an increase in the qAF8 levels was associated with a reduction in the odds of presenting with HRF, iRORA or cRORA (OR 0.91 [95% CI 0.85–0.97] per 10% increase in qAF8; $p = 0.006$). An increasing drusen volume in those without cRORA was associated with an increase in the odds of presenting with HRF, iRORA or cRORA (OR 2.00 [95% CI 1.47–2.73] per 0.1 unit increase in drusen volume; $p < 0.001$).

4. Discussion

Our study shows that the qAF8 levels are variable among the controls [2,30,31].

The mean qAF8 levels significantly decreased in AMD eyes (all AMD categories taken together) compared with the age-matched controls. The decrease in qAF8 may be explained by a change in the configuration of bisretinoids produced in AMD that may be less autofluorescent [32]. Alternatively, we can infer that the photodegradation of bisretinoids may be more rapid than the accumulation of photooxidised bisretinoids in AMD [32].

We also observed that RPE cell dysfunction appears early in AMD. Each AMD category was associated with reduced qAF8 compared to the age matched controls, with no significant differences in the qAF8 levels between them. These observations reflect early RPE dysfunction. Exocytosis may play a role [33]. As more RPE cell death occurs, more and more new RPE cells may undergo dysfunction, so the mean qAF8 may not be different between the AMD groups [16].

We also found that decreased qAF8 values are associated with eyes at risk of continuing RPE cell death [34,35]. Bird hypothesised that the reduced ability of RPE cells to degrade the outer segments may reduce lipid availability to the photoreceptor cells to produce outer segments and the shortening of outer segments would follow [36]. In turn, there may be reduced shedding and decreased phagosomal load in the RPE and reduced lipofuscin formation.

The qAF8 levels did not vary significantly with the drusen volume in eyes with AMD [37]. Nonetheless, the drusen volume was found to be increased in people with HRF, iRORA and cRORA, in keeping with the fact that the RPE overlying the drusen tends to thin and disintegrate, leading to the appearance of HRF and consequent low qAF8 values [38].

Another hypothesis is that reduced qAF8 levels may suggest abnormal visual cycle as the spatial location of lipofuscin reflects those of the rods [2]. However, there was no statistical difference in the qAF8 levels between the AMD eyes with and without SDD. As SDDs are markers of a delayed rod intercept time, our findings do not suggest that rod dysfunction or changes in the visual cycle influences the qAF8 levels [39].

Although anti-lipofuscin interventions may be useful to reduce the toxicity of lipofuscin on the RPE cells, our study implies that very early intervention is required to prevent RPE dysfunction in people aged less than 70 years. After 70 years, there seems to be a normal decline in the RPE cell function based on the qAF8 values. Although late onset Stargardt disease can be confused with GA, there were no eyes with GA and high qAF8 in this study.

Our study has several strengths. Firstly, our study is the largest study on qAF8 in AMD to date, and we included a large proportion of eyes with SDD and a wide age-range, particularly above 70 years. Both the Beckman and multimodal classifications were used to validate our findings. The participants with early AMD were grouped separately, rather than together with the healthy cohort, to understand the differences between them and healthy eyes. The multimodal classification allowed for the deep phenotyping of AMD. Thus, we could evaluate the changes in qAF8 in eyes with SDD and varying degrees of RPE loss. Our study is strengthened by a uniform and validated methodology of image acquisition and analysis and the use of robust statistics with adjustment for age and gender.

However, our study has a few limitations. The intrinsic biologic limitations of qAF8 imaging cannot be ignored [16,40]. We had a large percentage of phakic eyes that were adjusted for the transmission factor, and this may have influenced the qAF8 levels. Additionally, we used only the middle qAF8 segments, which is not representative of the entire macula. Nonetheless, including other segments has its own limitations; therefore, the most validated measurement of qAF8 was used [41].

In conclusion, we show that there is a decline in qAF8 levels in eyes with AMD compared with age-matched controls. We have shown various lines of evidence that qAF8 is reduced during early AMD progression. The qAF8 levels seem related more to the loss of function and integrity of RPE cells rather than being due to abnormalities in the visual cycle.

Supplementary Materials: The following supporting information can be downloaded at: https://www.mdpi.com/article/10.3390/biomedicines11020560/s1, Figure S1. Flowchart showing the analysed participant cohort; Figure S2. Correlation between Age and qAF8 levels in AMD and age-matched controls; Table S1. Participant and imaging characteristics in AMD categories based on the Beckman and OCT classification; Table S2. Age, gender and lens status adjusted associations between stage of SDD, qAF8 levels and drusen volume on risk factors of RPE cell death in AMD eyes.

Author Contributions: Conceptualisation, S.C. and S.S.; methodology, S.C. and S.S.; formal analysis, S.G. and S.C.; investigation, S.C., M.K.G. and R.S.; data curation, S.C. and R.S.; writing—original draft preparation, S.C., S.G. and S.S.; writing—review and editing, G.J., V.C., S.S. and A.B.; supervision, S.S., V.C., G.J. and A.B.; funding acquisition, S.S. All authors have read and agreed to the published version of the manuscript.

Funding: This research was funded by Fight For Sight, UK, grant number 1905.

Institutional Review Board Statement: The study was conducted in accordance with the Declaration of Helsinki and approved by the London-Chelsea Research Ethics Committee, London (REC 19/LO/0931).

Informed Consent Statement: Informed consent was obtained from all subjects involved in the study. Written informed consent has been obtained from the patient(s) to publish this paper.

Data Availability Statement: The data are not publicly available due to confidentiality issues and will only be available on reasonable request from the corresponding author.

Acknowledgments: This study was supported by the NIHR Biomedical Research Centre at Moorfields Eye Hospital NHS Foundation Trust and UCL Institute of Ophthalmology and the NIHR Moorfields Clinical Research Facility. The views expressed are those of the author(s) and not necessarily those of the NHS, the NIHR or the Department of Health and Social Care.

Conflicts of Interest: The authors declare no conflict of interest. Victor Chong is an employee of Janssen R&D LLC, but this work is not related to his employment with Janssen and the manuscript is not endorsed by Janssen. No conflict related to this project. Sobha Sivaprasad has received funding/fees from Bayer, Novartis, Allergan, Roche, Boehringer Ingelheim, Optos, EyeBiotech, Biogen, and Apellis. No conflicts related to this project. The funders had no role in the design of the study; in the collection, analyses, or interpretation of data; in the writing of the manuscript; or in the decision to publish the results.

References

1. Okubo, A. The relationships of age changes in retinal pigment epithelium and Bruch's membrane. *Investig. Ophthalmol. Vis. Sci.* **1999**, *40*, 443–449. [PubMed]
2. Delori, F.C.; Goger, D.G.; Dorey, C.K. Age-related accumulation and spatial distribution of lipofuscin in RPE of normal subjects. *Investig. Ophthalmol. Vis. Sci.* **2001**, *42*, 1855–1866.
3. Moreno-García, A.; Kun, A.; Calero, O.; Medina, M.; Calero, M. An overview of the role of lipofuscin in age-related neurodegeneration. *Front. Neurosci.* **2018**, *12*, 464. [CrossRef]
4. Young, R.W. Pathophysiology of age-related macular degeneration. *Surv. Ophthalmol.* **1987**, *31*, 291–306. [CrossRef] [PubMed]
5. Delori, F.C.; Fleckner, M.R.; Goger, D.G.; Weiter, J.J.; Dorey, C.K. Autofluorescence distribution associated with drusen in age-related macular degeneration. *Investig. Ophthalmol. Vis. Sci.* **2000**, *41*, 496–504.
6. Pollreisz, A.; Messinger, J.D.; Sloan, K.R.; Mittermueller, T.J.; Weinhandl, A.S.; Benson, E.K.; Kidd, G.; Schmidt-Erfurth, U.; Curcio, C.A. Visualizing melanosomes, lipofuscin, and melanolipofuscin in human retinal pigment epithelium using serial block face scanning electron microscopy. *Exp. Eye Res.* **2018**, *166*, 131–139. [CrossRef] [PubMed]
7. Brunk, U.T.; Terman, A. Lipofuscin: Mechanisms of age-related accumulation and influence on cell function. *Free Radic. Biol. Med.* **2002**, *33*, 611–619. [CrossRef] [PubMed]
8. Holz, F.G. Patterns of increased in vivo fundus autofluorescence in the junctional zone of geographic atrophy of the retinal pig-ment epithelium associated with age-related macular degeneration. *Graefes Arch. Clin. Exp. Ophthalmol.* **1999**, *237*, 145–152. [CrossRef]
9. Sparrow, J.R.; Duncker, T. Fundus autofluorescence and RPE lipofuscin in age-related macular degeneration. *J. Clin. Med.* **2014**, *3*, 1302–1321. [CrossRef]
10. Sparrow, J.R.; Duncker, T.; Schuerch, K.; Paavo, M.; de Carvalho, J.R.L., Jr. Lessons learned from quantitative fundus autofluorescence. *Prog. Retin. Eye Res.* **2020**, *74*, 100774. [CrossRef]
11. Burke, T.R.; Duncker, T.; Woods, R.L.; Greenberg, J.P.; Zernant, J.; Tsang, S.H.; Smith, R.T.; Allikmets, R.; Sparrow, J.R.; Delori, F.C. Quantitative fundus autofluorescence in recessive Stargardt disease. *Investig. Ophthalmol. Vis. Sci.* **2014**, *55*, 2841–2852. [CrossRef]
12. Gliem, M.; Müller, P.L.; Finger, R.P.; McGuinness, M.B.; Holz, F.G.; Issa, P.C. Quantitative fundus autofluorescence in early and intermediate age-related macular degeneration. *JAMA Ophthalmol.* **2016**, *134*, 817–824. [CrossRef]
13. Greenberg, J.P.; Duncker, T.; Woods, R.L.; Smith, R.T.; Sparrow, J.R.; Delori, F.C. Quantitative fundus autofluorescence in healthy eyes. *Investig. Ophthalmol. Vis. Sci.* **2013**, *54*, 5684–5693. [CrossRef]
14. Reiter, G.S.; Schwarzenbacher, L.; Schartmüller, D.; Röggla, V.; Leydolt, C.; Menapace, R.; Schmidt-Erfurth, U.; Sacu, S. Influence of lens opacities and cataract severity on quantitative fundus autofluorescence as a secondary outcome of a randomized clinical trial. *Sci. Rep.* **2021**, *11*, 12685. [CrossRef]
15. Wang, Y.; Tran, T.; Firl, K.; Huang, N.; Yasin, O.; van Kuijk, F.J.; Montezuma, S.R. Quantitative fundus autofluorescence in smokers compared to non-smokers. *Exp. Eye Res.* **2019**, *184*, 48–55. [CrossRef]
16. Ach, T.; Huisingh, C.; McGwin, G., Jr.; Messinger, J.D.; Zhang, T.; Bentley, M.J.; Gutierrez, D.B.; Ablonczy, Z.; Smith, R.T.; Sloan, K.R.; et al. Quantitative autofluorescence and cell density maps of the human retinal pigment epithelium. *Investig. Ophthalmol. Vis. Sci.* **2014**, *55*, 4832–4841. [CrossRef]

17. Mata, N.L.; Lichter, J.B.; Vogel, R.; Han, Y.; Bui, T.V.; Singerman, L.J. Investigation of oral fenretinide for treatment of geographic atrophy in age-related macular degeneration. *Retina* **2013**, *33*, 498–507. [CrossRef]
18. Petrukhin, K. Pharmacological inhibition of lipofuscin accumulation in the retina as a therapeutic strategy for dry AMD treatment. *Drug Discov. Today Ther. Strat.* **2013**, *10*, e11–e20. [CrossRef]
19. Rosenfeld, P.J. Emixustat hydrochloride for geographic atrophy secondary to age-related macular degeneration: A randomized clinical trial. *Ophthalmology* **2018**, *125*, 1556–1567. [CrossRef]
20. Sadda, S.R. Consensus definition for atrophy associated with age-related macular degeneration on OCT: Classification of atrophy report 3. *Ophthalmology* **2018**, *125*, 537–548. [CrossRef]
21. von der Emde, L. Natural history of quantitative autofluorescence in intermediate age-related macular de-generation. *Retina* **2021**, *41*, 694–700. [CrossRef]
22. van de Kraats, J.; van Norren, D. Optical density of the aging human ocular media in the visible and the UV. *J. Opt. Soc. Am. A. Opt. Image Sci. Vis.* **2007**, *24*, 1842–1857. [CrossRef]
23. Ferris, F.L., 3rd; Wilkinson, C.P.; Bird, A.; Chakravarthy, U.; Chew, E.; Csaky, K.; Sadda, S.R. Clinical classification of age-related macular degeneration. *Ophthalmology* **2013**, *120*, 844–851. [CrossRef]
24. Spaide, R.F.; Ooto, S.; Curcio, C.A. Subretinal drusenoid deposits AKA pseudodrusen. *Surv. Ophthalmol.* **2018**, *63*, 782–815. [CrossRef]
25. Kim, D.Y. Comparison of single drusen size on color fundus photography and spectral-domain optical coherence tomography. *Retina* **2021**, *41*, 1715–1722. [CrossRef]
26. Spaide, R.F.; Curcio, C.A. Drusen characterization with multimodal imaging. *Retina* **2010**, *30*, 1441–1454. [CrossRef]
27. Hardin, J.W.; Hilbe, J.M. *Generalized Estimating Equations*, 1st ed.; Chapman and Hall/CRC: Boca Raton, FL, USA, 2002.
28. Liang, K.-Y.; Zeger, S.L. Longitudinal data analysis using generalized linear models. *Biometrika* **1986**, *73*, 13–22. [CrossRef]
29. Shih, J.H.; Fay, M.P. Pearson's chi-square test and rank correlation inferences for clustered data. *Biometrics* **2017**, *73*, 822–834. [CrossRef]
30. Orellana-Rios, J.; Yokoyama, S.; Agee, J.M.; Challa, N.; Freund, K.B.; Yannuzzi, L.A.; Smith, R.T. Quantitative fundus autofluorescence in non-neovascular age-related macular degeneration. *Ophthalmic Surg. Lasers Imaging Retin.* **2018**, *49*, S34–S42. [CrossRef]
31. Armenti, S.T.; Greenberg, J.P.; Smith, R.T. Quantitative fundus autofluorescence for the evaluation of retinal diseases. *J. Vis. Exp.* **2016**, *11*, 53577. [CrossRef]
32. Sparrow, J.R.; Gregory-Roberts, E.; Yamamoto, K.; Blonska, A.; Ghosh, S.K.; Ueda, K.; Zhou, J. The bisretinoids of retinal pigment epithelium. *Prog. Retin. Eye Res.* **2012**, *31*, 121–135. [CrossRef]
33. Ferrington, D.A.; Sinha, D.; Kaarniranta, K. Defects in retinal pigment epithelial cell proteolysis and the pathology associated with age-related macular degeneration. *Prog. Retin. Eye Res.* **2016**, *51*, 69–89. [CrossRef] [PubMed]
34. Jaffe, G.J. Imaging features associated with progression to geographic atrophy in age-related macular degeneration: Classification of atrophy meeting report 5. *Ophthalmol. Retina* **2021**, *5*, 855–867. [CrossRef] [PubMed]
35. Balaratnasingam, C. Cuticular drusen: Clinical phenotypes and natural history defined using multimodal imaging. *Ophthalmology* **2018**, *125*, 100–118. [CrossRef]
36. Bird, A. Role of retinal pigment epithelium in age-related macular disease: A systematic review. *Br. J. Ophthalmol.* **2021**, *105*, 1469–1474. [CrossRef]
37. Reiter, G.S.; Told, R.; Schlanitz, F.G.; Bogunovic, H.; Baumann, L.; Sacu, S.; Schmidt-Erfurth, U.; Pollreisz, A. Impact of drusen volume on quantitative fundus autofluorescence in early and intermediate age-related macular degeneration. *Investig. Opthalmol. Vis. Sci.* **2019**, *60*, 1937–1942. [CrossRef] [PubMed]
38. Schmidt-Erfurth, U.; Bogunovic, H.; Grechenig, C.; Bui, P.; Fabianska, M.; Waldstein, S.; Reiter, G.S. Role of deep learning—Quantified hyperreflective foci for the prediction of geographic atrophy progression. *Am. J. Ophthalmol.* **2020**, *216*, 257–270. [CrossRef]
39. Owsley, C.; Clark, M.E.; McGwin, G. Natural history of rod-mediated dark adaptation over 2 years in intermediate age-related macular degeneration. *Transl. Vis. Sci. Technol.* **2017**, *6*, 15. [CrossRef] [PubMed]
40. Ablonczy, Z.; Higbee, D.; Anderson, D.M.; Dahrouj, M.; Grey, A.; Gutierrez, D.; Koutalos, Y.; Schey, K.L.; Hanneken, A.; Crouch, R.K. Lack of correlation between the spatial distribution of A2E and lipofuscin fluorescence in the human retinal pigment epithelium. *Investig. Ophthalmol. Vis. Sci.* **2013**, *54*, 5535–5542. [CrossRef]
41. Kleefeldt, N.; Bermond, K.; Tarau, I.-S.; Hillenkamp, J.; Berlin, A.; Sloan, K.R.; Ach, T. Quantitative fundus autofluorescence: Advanced analysis tools. *Transl. Vis. Sci. Technol.* **2020**, *9*, 2. [CrossRef]

biomedicines

MDPI

Article

Suppression of Age-Related Macular Degeneration-like Pathology by c-Jun N-Terminal Kinase Inhibitor IQ-1S

Anna A. Zhdankina [1,*], Dmitry I. Tikhonov [1], Sergey V. Logvinov [1], Mark B. Plotnikov [2,3], Andrei I. Khlebnikov [4] and Nataliya G. Kolosova [5]

1 Department of Histology, Embryology and Cytology, Siberian State Medical University, 634055 Tomsk, Russia
2 Department of Pharmacology, Goldberg Research Institute of Pharmacology and Regenerative Medicine, Tomsk NRMC, 3 Lenin Ave., 634028 Tomsk, Russia
3 Radioiphysical Faculty, National Research Tomsk State University, 634050 Tomsk, Russia
4 Kizhner Research Center, Tomsk Polytechnic University, 30 Lenin Ave., 634050 Tomsk, Russia
5 Federal Research Center Institute of Cytology and Genetics SB RAS, Pr. Lavrentiev, 10, 630090 Novosibirsk, Russia
* Correspondence: annazhdank@yandex.ru

Abstract: Age-related macular degeneration (AMD) is the leading cause of irreversible visual impairment worldwide. The development of AMD is associated with inflammation, oxidative stress, and progressive proteostasis imbalance, in the regulation of which c-Jun N-terminal kinases (JNK) play a crucial role. JNK inhibition is discussed as an alternative way for prevention and treatment of AMD and other neurodegenerative diseases. Here we assess the retinoprotective potential of the recently synthesized JNK inhibitor 11*H*-indeno[1,2-*b*]quinoxalin-11-one oxime sodium salt (IQ-1S) using senescence-accelerated OXYS rats as a model of AMD. The treatment with IQ-1S (50 mg/kg body weight intragastric) during the period of active disease development (from 4.5 to 6 months of age) improved some (but not all) histological abnormalities associated with retinopathy. IQ-1S improved blood circulation, increased the functional activity of the retinal pigment epithelium, reduced the VEGF expression in the endothelial cells, and increased the expression of PEDF in the neuroretina. The result was a decrease in the degeneration of photoreceptors and neurons of the inner layers. IQ-1S significantly improved the retinal ultrastructure and increased the number of mitochondria, which were significantly reduced in the neuroretina of OXYS rats compared to Wistar rats. It seems probable that using IQ-1S can be a good prophylactic strategy to treat AMD.

Keywords: age-related macular degeneration; c-Jun N-Terminal Kinase Inhibitor IQ-1S; OXYS rats

Citation: Zhdankina, A.A.; Tikhonov, D.I.; Logvinov, S.V.; Plotnikov, M.B.; Khlebnikov, A.I.; Kolosova, N.G. Suppression of Age-Related Macular Degeneration-like Pathology by c-Jun N-Terminal Kinase Inhibitor IQ-1S. *Biomedicines* **2023**, *11*, 395. https://doi.org/10.3390/biomedicines11020395

Academic Editor: James A. Marrs

Received: 28 November 2022
Revised: 21 January 2023
Accepted: 25 January 2023
Published: 29 January 2023

1. Introduction

Age-related macular degeneration (AMD) is the leading cause of irreversible visual impairment and blindness in industrialized countries, and is defined as a chronic, multifactorial, and progressive central retinal disease. The prevalence of AMD is increasing dramatically as the proportion of the elderly in the population continues to rise [1]. AMD is a multifactorial disease involving a complex interplay of genetic, environmental, metabolic, and functional factors. Clinically, AMD is classified into dry (atrophic) AMD and wet (exudative) AMD depending on the presence of choroidal neovascularization. Most AMD starts as the dry type, and in 10–20% of individuals it progresses to the wet type. Although the introduction of anti-angiogenesis therapy has helped to prevent blindness and restore vision in wet AMD, there remains no effective treatment for the patients with dry (~90% of all cases) AMD [2]. The dry form is characterized by progressive degeneration of the retinal pigment epithelium (RPE) and photoreceptor cells that is due to age-related changes in the retina [3,4]. The etiology of AMD remains largely unknown, but it has been suggested that an equilibrium between proangiogenic vascular endothelial growth factor (VEGF) and antiangiogenic neurotrophic pigment epithelium-derived factor (PEDF) is important for regulation of retinal physiology and for pathophysiology, including late AMD [5–7]. Anti-VEGF treatment has turned out to be a good way to control neovascular wet AMD more effectively, but there is no effective treatment for the dry form.

It is also obvious that the development of this complex degenerative disease is associated with inflammation, oxidative stress, and progressive proteostasis imbalance in the endoplasmic reticulum (ER stress). Representatives of the family of stress-induced, mitogen-activated protein kinases (MAPs) play an important role in the processes of inflammation and the production of cytokines, apoptosis, neurodegeneration, proliferation, and the differentiation of cells. Among them are c-Jun N-terminal kinases (JNK), which modulate various cellular processes, including cell proliferation, apoptosis, autophagy, and inflammation. Three different genes encode three isoforms of the protein: JNK1, JNK2, and JNK3. JNK1 and JNK2 are expressed ubiquitously throughout the body, while JNK3 expression is seen mainly in the brain, heart, and testicles [8]. There is strong evidence that alterations in JNK signaling play a crucial role in the pathogenesis of neurodegenerative diseases, because the sustained activation of JNK leads to synaptic dysfunction and even neuronal apoptosis. Violations of signaling in JNK-controlled pathways are observed in the development of various neurodegenerative diseases [9]. As studies on various retinal cell cultures and animal models indicate, JNK signaling may contribute to the pathogenesis of AMD [10]. This is indicated by the fact that mice lacking JNK1 exhibit decreased inflammation, reduced choroidal neovascularization (CNV), lower levels of choroidal VEGF, and impaired choroidal macrophage recruitment in a murine model of wet AMD [11]. Wherein hypoxia-induced JNK1 activation promotes retinal VEGF production and pathological angiogenesis in a murine model of retinopathy of prematurity [12].

It is natural that pharmacological JNK inhibition is discussed as an alternative avenue for prevention and treatment of AMD as well as other neurodegenerative diseases [13]. To date, a number of non-protein synthetic JNK inhibitors have been described, including SP600125, AS601245, IQ-1S, SR-3306, and SU3327, as well as protein and non-protein molecules that inhibit JNK signaling [7,14–19]. Some of them have demonstrated neuroprotective activity in animal models of stroke [20–24]. Among them is the recently synthesized JNK inhibitor 11*H*-indeno[1,2-*b*]quinoxalin-11-one oxime sodium salt (IQ-1S), whose ability to inhibit JNK has been shown in silico [19,25,26], and neuroprotective properties were demonstrated in models of cerebral ischemia [20,27,28].

The aim of this study was to evaluate the retinoprotective potential of IQ-1S using senescence-accelerated OXYS rats, characterized by developing of a retinopathy that by the clinical, morphological, and ultrastructural signs corresponds to AMD in humans [29–31]. The first clinical manifestations of retinopathy are detected during ophthalmoscopic examinations in ~20% of OXYS rats at 5–6 weeks of age, and at the age of 3–4 months signs of retinopathy are recorded in all animals. Pathological changes progress and reach pronounced stages, suggesting loss or significant deterioration in visual acuity, by the age of 14–18 months. Retinopathy that develops in OXYS rats corresponds to the "dry" form of AMD, and is manifested by the signs for this disease: dystrophic changes and thinning of the retina, impaired microcirculation in the choroid, changes in neurotrophic supply, accumulation of lipofuscin and amyloid β, and structural characteristics of AMD [7,14,32]. At the same time, neovascularization develops with age in some ($\sim$10–20%) OXYS rats, like in humans. To assess the therapeutic potential of IQ-1S, we used OXYS rats aged 4.5 months. As we showed earlier, at this age OXYS rats exhibit the pronounced signs corresponding to the first stage of the disease (according to the Age-Related Eye Disease Study grade protocol—http://eyephoto.ophth.wisc.edu (acessed on 28 November 2022)): appearance of drusen and other pathological changes in the retinal pigmented epithelium (RPE), and partial atrophy of the choroid capillary layer [33]. Here we examined the effect of IQ-1S on the AMD-like pathology development based on a structural analysis of the retina and expression of VEGF and PEDF.

2. Materials and Methods

2.1. Animals

The OXYS rat strain was developed at the Institute of Cytology and Genetics (ICG), SB RAS (Novosibirsk, Russia), from a Wistar stock as described earlier [34]. Males of 4-month-old OXYS rats, and age-matched Wistar rats, were obtained from the Breeding Experimental Animal Laboratory of the ICG SB RAS (Novosibirsk, Russia). The animals were kept under standard laboratory conditions (22 ± 2 °C, 60% relative humidity, and 12 h light/12 h dark cycle) and had ad libitum access to standard rodent feed (PK-120-1, Laboratorsnab, Ltd., Moscow, Russia) and water. To assess the influence of oral IQ-1S administration on the progression of the AMD-like pathology, we randomly assigned 4-month-old male OXYS rats to one of two groups (10 rats per group). One group was given intragastric administration of 50 mg of IQ-1S per kilogram of body weight in 2 mL of 1% starch mucus for 45 days (from age 4.5 to 6 months). The second group of OXYS rats received only starch mucus. As controls, we also used a group of 10 Wistar rats (parent strain, healthy rats). All experimental procedures were in compliance with the European Communities Council Directive of 24 November 1986 (86/609/EEC). The animal study was approved by the Commission on Bioethics of the Siberian State Medical University (Protocol No. 4 008/4/06/2022 dated 20 June 2022). Every effort was made to minimize the number of animals used and their discomfort. The IQ-1S-treated and untreated OXYS rats and Wistar rats were euthanized by an overdose of isoflurane inhalation anesthesia; the posterior wall of both eyes were carefully removed.

2.2. IQ-1S

The sodium salt of 11*H*-indeno[1,2-*b*]quinoxalin-11-one oxime (IQ-1S) (M314 series) was synthesized as described previously [19]. The chemical structure of IQ-1S was confirmed by the methods of mass spectrometry and nuclear magnetic resonance; sample purity was 99.9%. To prepare the IQ-1S emulsion, a weighted amount of IQ-1S powder, corresponding to the proper dose for the animal, was aseptically pounded with a pestle with 20 µL of Tween 80; 2.0 mL of physiologic sodium chloride solution was then added to create a suspension.

2.3. Histological Examination

For histological and immunohistochemical assays, the samples were fixed in 10% neutral formaldehyde in 0.1 mol/L phosphate buffer (pH 7.4) and embedded in paraffin according to the standard method [35]. Serial frontal sections (4 to 5 µm thick) were made, stained with H&E, and examined with a photomicroscope (Axiostar Plus, Carl Zeiss, Germany). Morphometric parameters were measured by quantitative image analysis performed with Axiovision software, version 4.8 (Zeiss, Thornwood, NY, USA). Evaluation was performed by examining five sections of the retina for each animal at a magnification of 10 × 100, using a frame area of 900 μm^2. The specific area of the choroidal vessels (open, with stasis, aggregation of blood cells, or thrombosis), and the specific area of the RPE layer, were measured, and their ratio to the total area of the choroid or retina, respectively, was calculated. Separately, the number of ganglionic neurons with central and total chromatolysis, and nuclear pyknosis per 200 corresponding cells of the layer, was counted. The percentage of photoreceptors with nuclear pyknosis per 1000 photoreceptors, the number of layers in 10 fields of view of each retinal slice, and the distribution density of the photoreceptor nuclei in the eyepiece frame of 900 μm^2 were calculated, these were then recalculated for an area of 1 mm^2 of the slice; the percentage of radial glial cells and neurons in the inner nuclear and ganglionic layers per 200 corresponding retinal cells was calculated.

2.4. Immunohistochemical Examination (Retinal Immunohistochemistry Staining)

Slides containing intact tissue were placed in a thermostat at 37 °C for 30 minutes. The slide was then simultaneously deparaffinized by placing xylene-xylene-100% ethanol-100%

ethanol-95% ethanol-90% ethanol-80% ethanol-70% ethanol. After rinsing, the slides were immersed in 3% H_2O_2 for 10 min. Then buffer with citric acid was added to the glass slide and heated to boiling. The sections were blocked with 5% BSA and then incubated at 37 °C for half an hour. The sections were then incubated with primary antibodies to VEGF (Fine Biotech Co., Ltd., Wuhan, China), PEDF (Abcam, Waltham, MA, USA), and p53 (Novus Biologicals, Centennial, CA, USA), colored at 4 °C overnight. Then, after incubation for 30 min at +25 °C with the secondary antibody, the sections were washed in two portions of the buffer solution. To detect staining, sections were incubated with an aqueous solution of chromogen (AEC Chromogen/Substrate Kit, Novus Biologicals, USA) for 10 min at +25 °C, after which they were washed with two portions of tap water and counterstained with Mayer's hematoxylin. The immunohistochemical reaction was assessed as positive when reddish-brown staining was detected in the cytoplasm or on the membrane of the studied cells. Five rats were used and five sections of each rat were counted. The amount of all immunopositive structures as a percentage of the total number of structures in the retinal section was calculated.

2.5. Electron Microscopic Examination

Retinal samples of untreated and IQ-1S treated OXYS rats and Wistar rats (n = 5 per group) were fixed with 2.5% glutaraldehyde in 0.1 M sodium cacodylate buffer (pH 7.2) for 1 h, washed with 0.1 M sodium cacodylate buffer, and post-fixed in 1% osmium tetroxide in the same buffer for 1 h. After that, the samples were washed with water and incubated in a 1% aqueous solution of uranyl acetate in the dark at RT for 1 h. The samples were then dehydrated using a graduated series of ethanol and acetone mixtures, and embedded with a mixture of epon-araldite resins. First, semi-thin sections, 1 µm thick, were made on an ultratome and stained with toluidine blue, then ultra-thin sections were made. Ultra-thin sections were stained with uranyl acetate and lead citrate, and then examined under a transmission electron microscope (JEM 100 SX; Jeol, Tokyo, Japan) at the Interdepartmental Joint Center for Microscopic Analysis of Biological Objects, Institute of Cytology and Genetics, Siberian State University, Department of the Russian Academy of Sciences. On electron micrographs of the associative and ganglionic layers of the retina (45 photos per group of animals), all organelles located in these areas were stained using the software Adobe Photoshop. For each photograph, the following parameter was determined: the specific total area of each type of organelle located in the electron-transparent areas of neurons.

2.6. Statistics

Statistical analysis was carried out using the Statistica 10 software package (Statsoft, St Tulsa, OK, USA), implementing the methods of variation statistics. To assess the significance of differences when comparing mean values, the nonparametric Mann–Whitney test was used. The data are presented as median and interquartile range 25th and 75th percentiles (Me (Q25%–Q75%)) or as the mean± standard deviation of the mean (SD). Differences were considered statistically significant at $p < 0.05$.

3. Results

3.1. IQ-1S Attenuates Pathological Changes in the Retina of OXYS Rats

Consistent with our previous observations, by the age of 6 months there were alterations in all layers of the untreated OXYS rats' retinas. Thus, aberrations of RPE cells flattening, with a variable size and shape of their nuclei, typical to AMD, as well as some decrease in the thickness of the photoreceptor layer were identified (Table 1). In the RPE cells, we found pathological changes in the ultrastructure typical for the OXYS rats: disintegration of microvilli, disappearance of basal striation, accumulation of a large number of phagolysosomes, lipofuscin, and the presence of individual cells with marginal condensation of chromatin in the nucleus, which is one of the signs of the development of apoptosis (Figure 1).

Table 1. Effects of IQ-1S on the morphometric parameters of the Retina of OXYS Rats.

Parameters	Wistar	OXYS	OXYS + IQ-1S
Specific area of Open choriocapillaris, %	32.9 (24.6; 39.0)	12,1 (10.4; 15.0) *	20.0 (15.6; 20.8) *+
Specific area of choriocapillaris with stasis and thrombosis, %	1.71 (1.26; 1.86)	22.23 (15.59; 20.81) *	10.0 (6.3; 11.0) *+
Specific area of the layer RPE, %	6.57 (6.28; 7.30)	3,12 (2.99; 3.17) *	4.83 (4.53; 5.19) *+
Photoreceptors with nuclear pyknosis, %	0.40 (0.32; 0.40)	23.5 (17.5; 24.3) *	8.3 (7.8; 8.8) *+
Number of rows of photoreceptor nuclei	12.0 (11.3; 13.0)	10.0 (10.3; 11.8) *	10.0 (9.0; 10.8) *
Numerical density of photoreceptors, nuclei per mm^2	39,505 (35,022; 40,943)	24,935 (24,935; 26,485) *	32,958 (26,873; 33,918) *+
p53 immunopositive neurons in the ganglion layer, %	0	6.00 (3.42; 8.2) *	1.67 (1.00; 2.00) *+
p53 immunopositive neurons in the inner nuclear layer, %	0	4.47 (4.0; 7.0)	2.00 (1.94; 2.6)
p53 immunopositive neurons in the outer nuclear layer, %	0	0	0

*—significant differences ($p < 0.05$) between the strains; +—a significant effect ($p < 0.05$) of IQ-1S within the strain. Data are presented as median and interquartile range—25th and 75th percentiles (Me (Q25%–Q75%)). IQ-1S was given at 50 mg per day from 4.5 to 6 months.

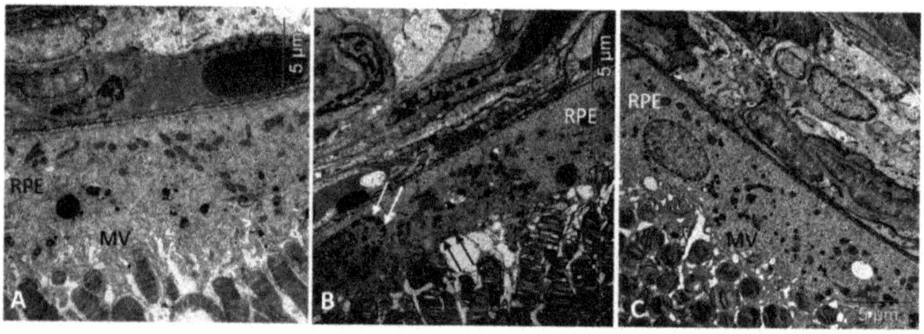

Figure 1. Treatment with IQ-1S improved the structural and functional parameters of RPE in 6-month-old OXYS rats. Electron micrographs show examples of normal ultrastructural signs of RPE in Wistar rats (**A**), untreated (**B**), and IQ-1S treated (**C**) OXYS rats. RPE images from untreated OXYS rats (**B**) show evidence of RPE cell disorganization-nuclear chromatin condensation (arrow) and microvilli destruction (black arrows) in OXYS rats. RPE cells from OXYS rats treated with IQ-1S (**C**) have normal nuclear and cytoplasmic ultrastructure. The presence of microvilli (MV) on their surface indicates that the cells are functionally active.

The degree of damage to photoreceptor cells in the untreated OXYS rats was significant. Photoreceptor cells died mainly due to necrosis, with the development of nuclear pyknosis, perikaryon edema, cell lysis, and proliferation of scleral processes in Muller cells, as evidenced by a significant decrease in the density of the photoreceptor cell nuclei and an increase in the percentage of pyknosis among this population (Figure 2, Table 1).

The degree of damage to neurons in the inner layers of the retinas of the OXYS rats was lower than in the photoreceptor layer. At the same time, both in the inner nuclear and ganglionic layers of the retinas of the OXYS rats, the frequency of apoptosis was higher than in the Wistar rats, as evidenced by an increase in cells with signs of pyknosis. Increased apoptosis in ganglionic neurons, and in the inner nuclear layer, was confirmed by the presence of cells positive for the apoptosis-related marker, p53 protein, which were not identified in the outer nuclear layer (Figure 3, Table 1).

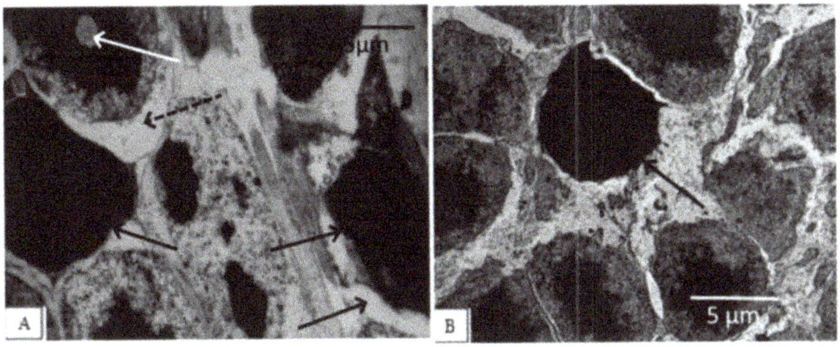

Figure 2. Outer nuclear layer of OXYS rats without correction (**A**) and on the background of IQ-1S treatment (**B**). In OXYS rats, pronounced changes in the nuclear part of the photoreceptors were observed: nuclear pyknosis (black arrows), perikaryon edema (dashed arrows), and nuclear lysis (white arrow). On the background of treatment with IQ-1S, the edema phenomena leveled out, a greater preservation of photoreceptor nuclei was observed. However, pyknosis phenomena took place.

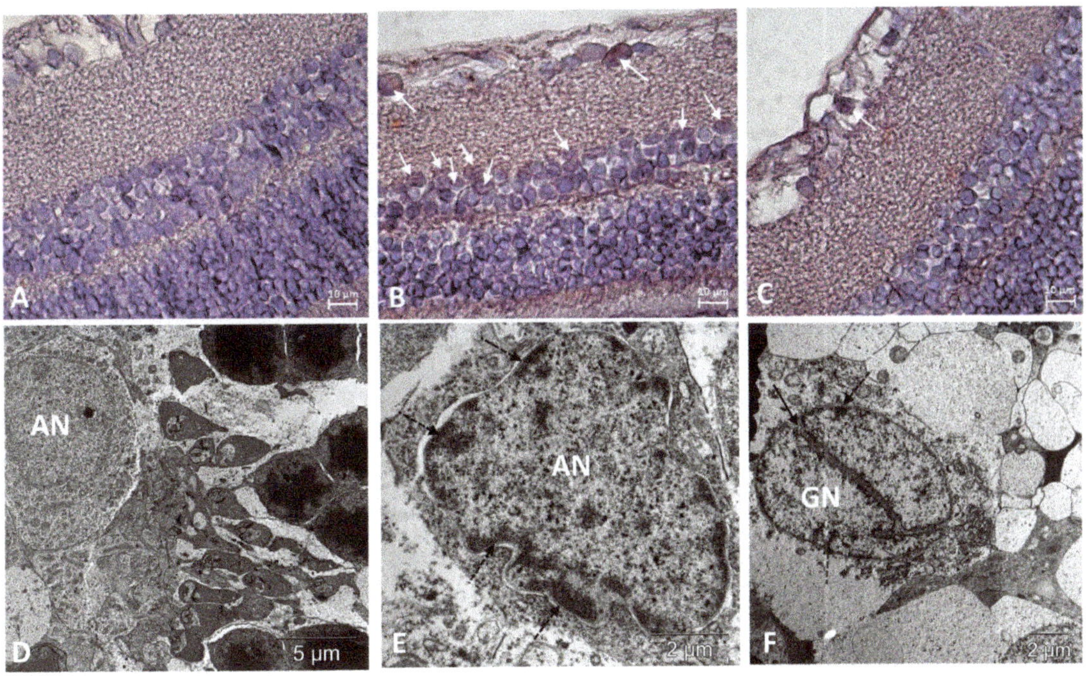

Figure 3. Immunostaining did not reveal p53-positive cells in the retinas of Wistar rats (**A**) and revealed them among neurons in the ganglionic and inner nuclear layers of the retina of OXYS rats (white arrows) (**B**). Treatment with IQ-1S significantly decreased the number of p53-positive cells in the retinas of OXYS rats (**C**). The electron micrographs show ultrastructural features of the normal structure of an associative neuron (AN) in Wistar rats (**D**) and examples of morphological manifestations of apoptosis in retinal neurons of OXYS rats: (**E**) nuclear invagination and marginal chromatin condensation (dashed arrows) in the associative neuron; (**F**) division of the nucleus of the ganglionic neuron (GN), marginal condensation of chromatin in the nucleus (dashed arrows). (**A–C**): representative images of immunohistochemistry. Additional staining with hematoxylin. (**D–F**): electron diffraction pattern.

These aberrations occur against the background of significant changes in the vessels of the choroid. Along with vessels with normal blood supply, we observed the appearance of a certain number of capillaries with signs of partial occlusion. In contrast to the choroids of Wistar rats, in the choroids of the OXYS rats, blood flow disturbances were revealed: the aggregation of blood cells, the stasis, and the thrombosis of small vessels. As a result, the specific area of open functional vessels in the OXYS rats was 2.7 times less, and the number of vessels with sludge, erythrocyte stasis, and thrombosis was 13 times greater than in the Wistar rats ($p < 0.05$). An ultrastructural study confirmed the development of obliteration of the choroidal vessels, and revealed the destruction of the connective tissue components of the Bruch's membrane in the form of an increase in osmiophilia and granular disintegration of collagen structures (Figure 4).

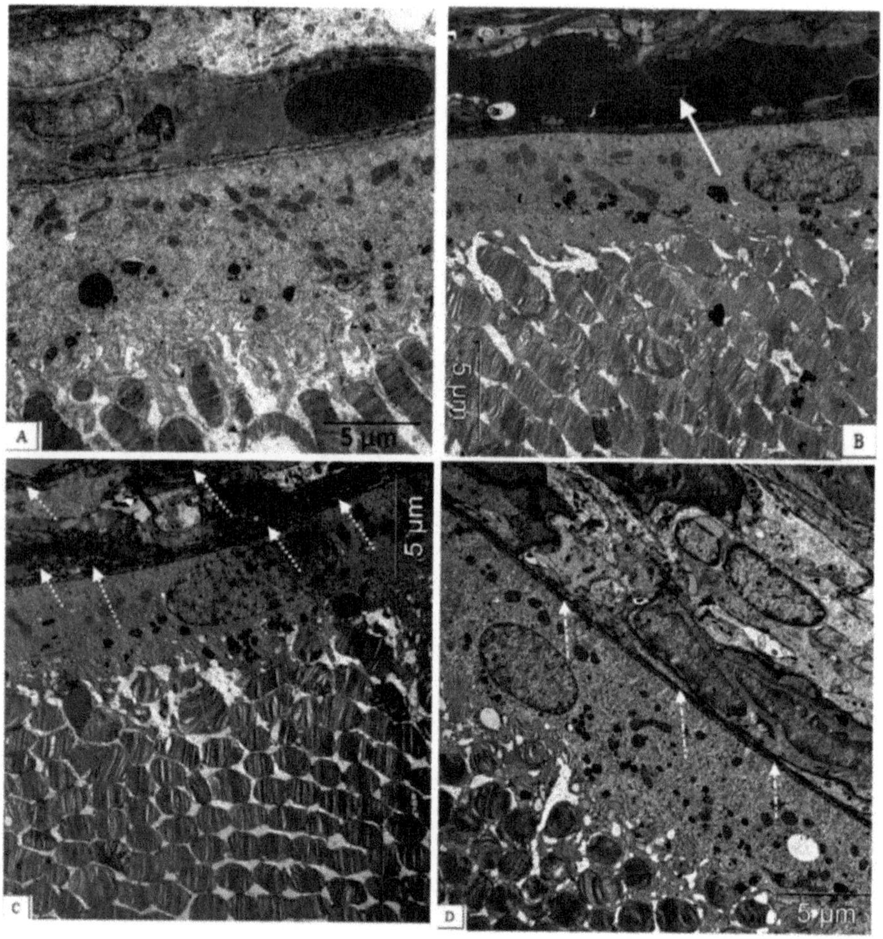

Figure 4. The electron micrographs show examples of normal ultrastructural features of the chorioretinal complex of a Wistar rat (**A**), untreated (**B**,**C**) and an IQ-1S-treated (**D**) OXYS rat. Stasis, sludge of the blood vessels of the OXYS rat's choroid (white arrow), destructive changes in the structure of the Bruch's membrane, and capillary endothelium (dashed arrow) are presented. During treatment with IQ-1S, the structure of the Bruch's membrane was intact, however, its thinning and increased osmiophilia were sometimes observed ((**D**), white dot-dotted arrow).

Treatment with IQ-1S significantly prevented thrombus formation in retinal vessels, which improved microcirculation and, consequently, led to the greater preservation of the RPE cells and photoreceptors. At the age of 6 months, the number of rows in the outer nuclear layer in the retinas of the OXYS rats (Table 1) was slightly, but significantly, less than in the Wistar rats. At the same time, the numerical density of photoreceptor nuclei was 1.5-fold less. The IQ-1S treatment did not affect the thickness of the photoreceptor layer and increased the density of photoreceptors; however, it remained somewhat lower than in Wistar rats. Importantly, the treatment with IQ-1S reduced the proportion of photoreceptors with signs of apoptosis approximately threefold, and also significantly reduced the proportion of the p53+ neurons in the ganglion layer and at the trend level in the inner nuclear layer. Additionally, IQ-1S improved the condition of the Bruch's membrane. However, electron microscopic examination showed that the majority of animals which received IQ-1S still had a decrease in the Bruch's membrane thickness and increased osmiophilia. Besides, RPE cells in IQ-1S-treated OXYS rats remained more flattened than in Wistar rats, but the presence of microvilli on their surface indicated that the cells were functionally active (Table 1, Figure 1).

Next, we assessed the influence of IQ-1S treatment on the ultrastructural state of the associative and ganglion neurons in the retinas of the OXYS rats. At the age of 6 months, substantial changes in the structural organization of cellular organelles were observed in the neurons of untreated OXYS rats in comparison with the Wistar rats (Figure 5, Table 2).

An electron microscopic study (Table 2) revealed a significant decrease in the number of mitochondria in the retinal neurons: the specific area of mitochondria in the cytoplasm of associative and ganglion neurons in 6-month-old OXYS rats was less than in the Wistar rats ($p < 0.01$ for both). Treatment with IQ-1S significantly increased this parameter ($p < 0.03$ for both cell types); however, the specific area of mitochondria in the cytoplasm of both associative and ganglionic neurons in the retinas of the OXYS rats remained significantly lower than in the Wistar rats ($p < 0.03$ and $p < 0.001$, respectively). Both associative and ganglion neurons of the OXYS rats have a smaller specific area of rough endoplasmic reticulum cisterns (ER, $p < 0.01$ for both). At the same time, mitochondria swell, the cristae are destroyed in them, the cisterns of the endoplasmic reticulum segregate, lose ribosomes, expand, which leads to the formation of vacuoles and is accompanied by an increase in the specific area of vacuoles in both associative and ganglionic neurons in the OXYS rats ($p < 0.02$ and $p < 0.01$, respectively). Treatment with IQ-1S increased the specific area of the rough ER in both types of neurons ($p < 0.03$), but only in associative neurons reached the level of the Wistar rats, while in ganglionic neurons it remained half as much ($p < 0.01$). Both associative and ganglion neurons in the retinas of the OXYS rats did not have significant differences in the specific area of the Golgi apparatus and lysosomes compared to the Wistar rats. Treatment with IQ-1S also did not affect the Golgi apparatus, however, it led to a decrease in the specific areas of vacuoles in both the associative and ganglion cells ($p < 0.03$ for both), but the indicator remained higher in the OXYS rats than in the Wistar rats.

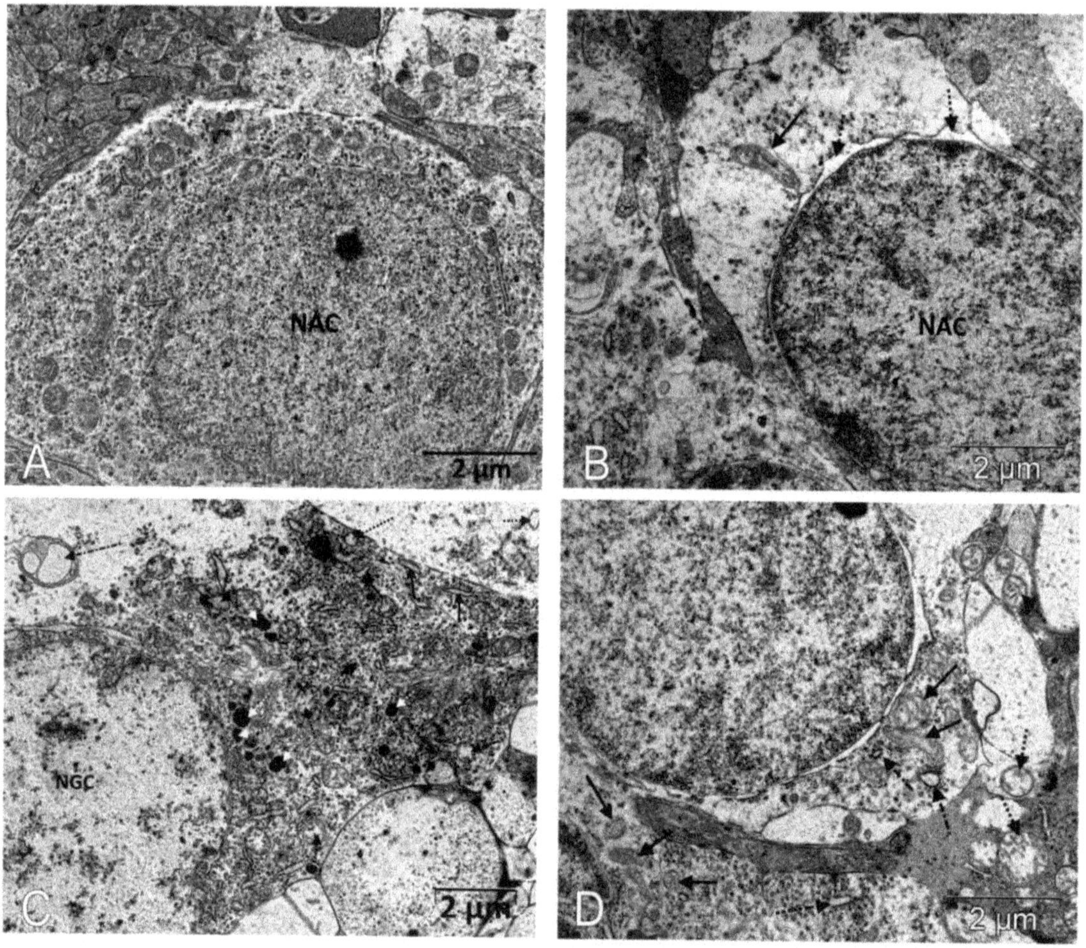

Figure 5. IQ-1S treatment improved the ultrastructure of neurons of the inner layers of the retinas in 6-month-old OXYS rats. The electron micrographs show examples of the ultrastructure of neurons of the inner layers of the retina of Wistar rats (**A**), untreated OXYS rats (**B**,**C**), and IQ-1S-treated OXYS rats (**D**). In the cytoplasm of the associative retinal neurons of untreated OXYS rats, there is an expansion of the perinuclear space (dashed arrows), destruction of most organelles, an increase in the size of preserved mitochondria, which have external signs of damage to the inner membrane of the cristae - local formation of electron-dense granules, tubular cristae (black arrow). (**B**). A focal destruction of organelles, a vacuolization, and a formation of multivesicular bodies (dot-dotted arrow) occur in most of the ganglion neurons; an expansion of granular ER cisterns (black arrows), an increase in the number of lysosomes (white dashed arrows), and a destruction of mitochondria are observed in the preserved parts of the cytoplasm (black dashed arrows) in the retinas of untreated OXYS rats (**C**). Treatment with IQ-1S leads to a greater preservation of organelles, including mitochondria (black arrows), but some of them increase in size; swelling and destruction of cristae are observed in them (dashed arrow). The cisterns of the granular ER are fragmented and expand, but do not lose their connection with the ribosomes (dot-dotted arrow) (**D**).

Table 2. Effects of IQ-1S on the specific areas of organelles in the cytoplasm of associative and ganglion neurons (according to electron micrographs).

Organelles	Type of neurons	Wistar	OXYS	OXYS + IQ-1S
Rough ER, %	Associative	20.41 (17.59; 29.72)	8.67 (3.89; 8.86) *	16.22 (12.69; 16.69) +
	Ganglion	43.85 (41.17; 59.28)	13.5 (13.09; 15.69) *	21.50 (16.24; 26.09) *+
Mitochondria, %	Associative	16.29 (16.13; 16.66)	3.51 (3.19; 3.79) *	9.23 (7.02; 14.43) *+
	Ganglion	16.98 (16.72; 17.49)	3.55 (2.26; 5.21) *	7.41 (6.53; 7.80) *+
Golgi apparatus, %	Associative	1.26 (1.18; 1.39)	0.64 (0.53; 0.83)	1.75 (1.16; 2.06)
	Ganglion	1.86 (1.43; 2.93)	1.66 (0.79; 1.77)	1.41 (1.09; 2.92)
Lysosomes, %	Associative	0.49 (0.35; 0.73)	0.31 (0.27; 0.37)	0.39 (0.02; 0.77)
	Ganglion	0.58 (0.55; 1.02)	0.56 (0.27; 0.80)	0.64 (0.02; 1.39)
Vacuoles, %	Associative	0.08 (0.06; 0.1)	2.40 (1.73; 2.4) *	1.00 (0.8; 1.09) *+
	Ganglion	0.19 (0.04; 0.36)	3.18 (2.37; 3.4) *	1.12 (0.98; 1.16) *+

*—significant differences ($p < 0.05$) between the strains; +—a significant effect ($p < 0.05$) of IQ-1S within the strain. Data are presented as median and interquartile range—25th and 75th percentiles (Me (Q25%–Q75%)). IQ-1S was given at 50 mg per day from 4.5 to 6 months.

3.2. *IQ-1S Improved the Expression of VEGF and PEDF in the Retina of OXYS Rats*

At the next stage, we evaluated the effect of IQ-1S on the expression of key regulators of angiogenesis, the main pro-angiogenic VEGF, and the most potent natural angiogenesis inhibitor and PEDF (Figure 6). VEGF immunohistochemical staining was weak in choroidal vessels in all animal groups, but was slightly elevated in the control OXYS rats and was not detectable in the IQ-1S treated OXYS rats. At the same time, VEGF immunoreactivity is pronounced in intraretinal vessels, where VEGF-positive cells were twice as many in the OXYS rats compared to the Wistar rats ($p < 0.001$). Treatment with IQ-1S significantly reduced this parameter ($p < 0.029$), but it still remained higher than in the Wistar rats ($p < 0.028$). VEGF immunoreactivity was also strongly detected in ganglion neurons, where the number of VEGF+ cells was twice as high in the OXYS rats as in the Wistar rats and was significantly reduced by IQ-1S treatment ($p < 0.01$ for both), and, as a result, did not differ from that in Wistar rats. At the same time, immunohistochemical staining did not reveal VEGF+ cells among associative neurons of all groups of rats.

We did not find PEDF+ cells in both the choroid and intraretinal vessels. PEDF immunoreactivity was clearly observed in the ganglion and association neuron cells in rat retinas, with PEDF+ cell density in both types of neurons being significantly lower in the OXYS rats than in the Wistar rats ($p < 0.01$). IQ-1S treatment increased the proportion of PEDF+ cells among both types of neurons ($p < 0.01$ for both).

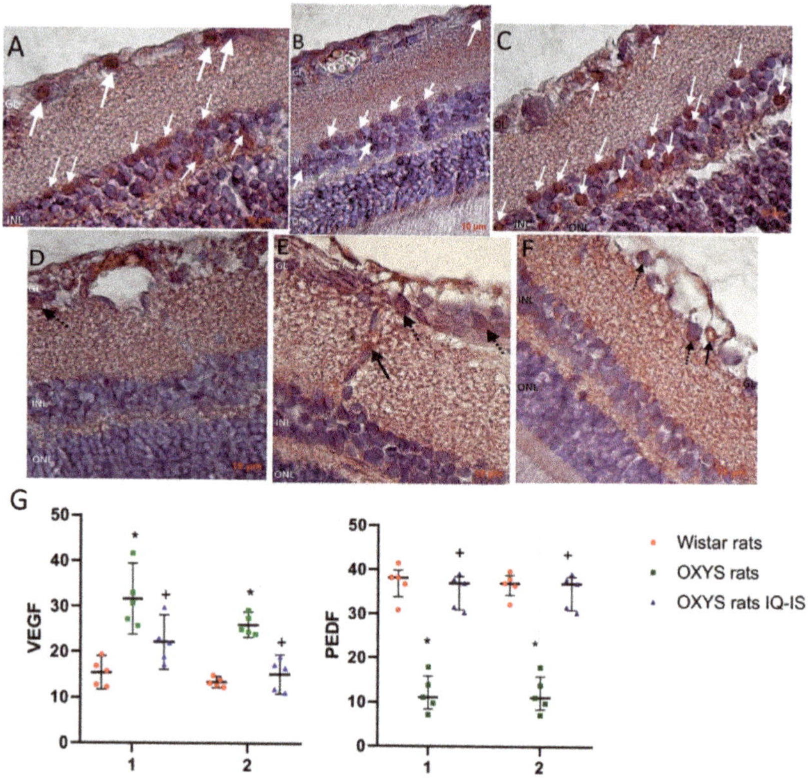

Figure 6. Detection and quantitative analysis of elements immunopositive to VEGF (**D–F**) and PEDF (**A–C**) in the retinas of Wistar rats (**A**,**D**), untreated OXYS rats (**B**,**E**), and IQ-1S-treated OXYS rats (**C**,**F**). Ganglionic neurons of the ganglionic layer (GL) and associative neurons of the inner nuclear layer (INL) were immunoreactive to PEDF in the neuroretina (white arrows). Vascular endothelium (black arrows) and ganglionic neurons (dashed arrows) were VEGF-immunopositive structures. IQ-1S treatment improved PEDF (**C**) and VEGF (**F**) expression in the retina of OXYS rats. Expression of VEGF in intraretinal vessels (1) and ganglion neurons (2) of Wistar rats, control rats, and OXYS rats treated with IQ-1S (**G**). The data are presented as the amount of VEGF- and PEDF-positive cells as a percentage of the total number of structures studied in the entire area of the retinal section (M $\pm$ SD).

4. Discussion

The aim of this study was to evaluate the retinoprotective potential of the JNK signaling pathway inhibitor IQ-1S using OXYS rats as an AMD model. As we have shown earlier, the clinical signs of AMD-like pathology against the background of structural and functional changes in RPE cells, and disorders of choroidal microcirculation, develop in 100% of the eyes of OXYS rats by the age of ~3–4 months [29,30,36]. Thus, it can be argued that at the time of the start of the treatment with the drug, the OXYS rats already had the corresponding clinical and pathomorphological signs of the disease. Here we have shown that at the age of 6 months, the main morphological signs of AMD-like pathology were expressed well in the retina of the OXYS rats from the control group. IQ-1S treatment started at 4 months of age significantly suppressed the progression of AMD-like pathology in the OXYS rats.

Retinopathy in the OXYS rats, consistent with dry AMD in humans, was characterized by progressive RPE degeneration and choroidal involution with secondary loss of photoreceptors. We have previously shown that destruction of RPE cells is the primary change in the development of retinopathy in OXYS rats, and begins as early as 20 days of age. It is manifested by a decrease in the area of RPE cells, an increase in the proportion of multinucleated cells, and a violation of their hexagonal shape. The progression of an AMD-like pathology in OXYS rats is accompanied by the excessive accumulation of beta-amyloid (Aβ) and lipofuscin in RPE cells, as well as a violation of cell morphology, their hypertrophy, and reactive gliosis [30,31,37]. In the present study, using light and electron microscopy, typical pathological changes in the morphology and ultrastructure of the RPE in 6-month-old OXYS rats were detected in the form of flattening of aberrational cells and destruction of microvilli. Treatment with IQ-1S significantly improved the structural and functional parameters of RPE. In treated animals, we observed a high preservation of cytoplasmic organelles, including the basal labyrinth and microvilli, which is a sign of the functional activity of the RPE. The specific area of the RPE was significantly higher than in the group without treatment, but did not reach the Wistar rat values.

There is growing evidence that changes in the choroid and choroidal microcirculation that occur with age may play a critical role in the pathogenesis of AMD, and localized choroidal dysfunction and abnormal hemodynamics are an important disease mechanism for AMD [38,39]. The manifestation of clinical signs of AMD in OXYS rats at the age of three months occurs against the background of age-related changes in blood flow in the choroidal vessels [30]. Here we have shown that pronounced signs of partial occlusion were observed in both choroidal and intraretinal vessels of 6-month-old OXYS rats. The violation of hemodynamics in the vessels of the choroid was accompanied by changes in the vascular wall, a fibrosis, and a violation of the structure of the Bruch's membrane, which was manifested by increased osmiophilia and granular breakdown of collagen structures. These data are consistent with the data of Zarbin M.A. [40]: under conditions of retinal aging, changes occur in the functioning of the Bruch's membrane proteins, accumulation of collagen glycosylation products, which leads to disruption of the RPE functioning, its destruction, and the death of photoreceptors. Treatment with IQ-1S significantly improved the state of retinal vessels, led to the preservation of the three-layer structure of the Bruch's membrane, which contributed to greater preservation of neurons in the retinas of OXYS rats. At the same time, some thinning of the membrane and an increase in its electron density were still observed, and blood flow disturbances in the choroid of OXYS rats persisted: the number of open choriocapillaries was reduced, and the number of choriocapillaries with signs of occlusion was increased compared to Wistar rats.

It is hypothesized that the JNK pathway may play a key role in the development of AMD due to its role in stress responses and involvement in apoptosis, inflammation, and VEGF production. JNK1 is a critical factor in hypoxia-induced retinal VEGF production and may contribute to hypoxia-induced pathological angiogenesis [12]. We have previously shown that the first signs of retinopathy in OXYS rats develop upon suppression of both proangiogenic VEGF and antiangiogenic neurotrophic PEDF proteins [36,41]. In the present study we revealed signs of higher expression of VEGF in intraretinal and choroidal vessels, as evidenced by an increase in the proportion of VEGF-positive cells. Nonetheless, we did not reveal signs of neovascularization in the retinas of 6-month-old untreated OXYS rats. Treatment with IQ-1S significantly decreased the proportion of the VEGF-positive endothelial cells, to the level of Wistar rats.

PEDF is multimodal and has neurotrophic, anti-angiogenic, anti-inflammatory, and antioxidant properties that can protect the cells of the inner retina and retinal ganglion cell layer from death caused by ischemia and cytotoxic agents [42]. Although there are a lot of data on the inhibition of VEGF production via the JNK-dependent pathway in the literature, we did not find data on the effect of JNK1 inhibitors on the expression of PEDF, as well as convincing data on its regulation via the JNK signaling pathway. At the same time, IQ-1S treatment not only reduced the number of VEGF-positive endothelial cells in retinal vessels in OXYS rats, but also increased the percentage of PEDF-positive associative and ganglion neurons, which was significantly reduced in untreated OXYS rats. Perhaps this effect of the drug is achieved by improving blood flow, reducing retinal tissue hypoxia, which also mediates the neuroprotective activity of IQ-1S.

Destructive changes in the RPE can lead to an increase in the death of photoreceptors, which according to morphological manifestations corresponds to necrosis, and is observed in the retina of untreated OXYS rats. At the same time, in the neurons of the inner layers of the retina of untreated OXYS rats, we observed an increase in perikaryon edema and a significant destruction of organelles. In addition, immunohistochemically in the neurons of this region of the retina of the control OXYS rats, we registered the activation of the p53-dependent apoptosis pathway. There are data in the literature on the role of JNK3 in p53 phosphorylation [43], which was indirectly confirmed in our study, since treatment with the JNK3 inhibitor IQ-1S significantly reduced the number of p53-positive associative and ganglion neurons in the retinas of OXYS rats, thus contributing to their survival.

There is growing evidence that mitochondrial dysfunction plays a critical role in the pathogenesis of AMD [44]. Mitochondrial dysfunction was considered the most likely cause of accelerated aging in OXYS rats. It was first identified in the liver [45], then in the muscles [46], the myocardium [47], and recently in brain structures [48]. Here we showed for the first time that the development of an AMD-like pathology in OXYS rats occurs against the background of disturbances in the ultrastructure of mitochondria and a significant decrease in their number in the neural retina: the specific area of mitochondria in the cytoplasm of associative and ganglionic neurons in 6-month-old OXYS rats was significantly less than in Wistar rats. IQ-1S treatment improved mitochondrial ultrastructure and increased their number.

Along with a decrease in the specific area of mitochondria in OXYS rats at the age of 6 months, a significant decrease in the specific area of granular ER was revealed, while cisternae preserved in the cytoplasm of neurons underwent pronounced ultrastructural changes: loss of ribosomes from the membrane surface, and segregation of cisternae with the formation of small cavities that expanded and served as a site for the formation of vacuoles. The observed changes are characteristic of a phenomenon called granular EPR stress [49]. At the same time, ER stress itself can stimulate cell death by modulating JNK activity [50]. IQ-1S reduced the manifestations of ER stress, which, together with the limitation of mitochondrial damage, led to a decrease in vacuolization and contributed to slowing down the destruction of neurons in the inner layers of the retina.

Collectively, our data support that the increase of the JNK-signaling pathway may play an important role in the pathophysiology of AMD. It seems probable that use of the JNK3 inhibitor IQ-1S can be a good prophylactic strategy to maintain retinal health and to treat AMD. However, for this, it is necessary to conduct a detailed study of the mechanisms of IQ-1S action.

Author Contributions: Conceptualization, A.A.Z., S.V.L., M.B.P. and N.G.K.; methodology, A.A.Z., S.V.L. and M.B.P.; formal analysis, A.A.Z., S.V.L., M.B.P. and N.G.K.; investigation, A.A.Z. and D.I.T.; resources, N.G.K., A.I.K. and M.B.P.; data curation, A.A.Z., S.V.L., M.B.P. and N.G.K.; writing-rough preparation, A.A.Z. and D.I.T.; writing-review and editing, A.A.Z., N.G.K., S.V.L. and M.B.P.; supervision, N.G.K., A.A.Z., S.V.L. and M.B.P.; project administration, A.A.Z.; acquisition of financing, A.A.Z. All authors have read and agreed to the published version of the manuscript.

Funding: The research was funded by the Russian Science Foundation, grant number 22-25-00686. Synthesis and identification of compound IQ-1S was supported by the Tomsk Polytechnic University Development Program.

Institutional Review Board Statement: The study was conducted according to Directive 2010/63/EU of the European Parliament and of the European Council of 22 September 2010 and was approved by the Commission on Bioethics of the Siberian State Medical University (Protocol No. 4 008/4/06/2022 dated 20 June 2022). Tomsk, Russia.

Informed Consent Statement: Not applicable.

Data Availability Statement: Raw data are available from the corresponding author upon request.

Acknowledgments: The animals were kindly provided by the Breeding Experimental Animal Laboratory of the ICG SB RAS (Novosibirsk, Russia).

Conflicts of Interest: The authors declare no conflict of interest.

References

1. Wong, W.L.; Su, X.; Li, X.; Cheung, C.M.; Klein, R.; Cheng, C.Y.; Wong, T.Y. Global prevalence of age-related macular degeneration and disease burden projection for 2020 and 2040: A systematic review and meta-analysis. *Lancet Glob. Health* **2014**, *2*, e106–e116. [CrossRef] [PubMed]
2. Gil-Martínez, M.; Santos-Ramos, P.; Fernández-Rodríguez, M.; Abraldes, M.J.; Rodríguez-Cid, M.J.; Santiago-Varela, M.; Fernández-Ferreiro, A.; Gómez-Ulla, F. Pharmacological Advances in the Treatment of Age-related Macular Degeneration. *Curr. Med. Chem.* **2020**, *27*, 583–598. [CrossRef] [PubMed]
3. Blasiak, J.; Sobczuk, P.; Pawlowska, E.; Kaarniranta, K. Interplay between aging and other factors of the pathogenesis of age-related macular degeneration. *Ageing Res. Rev.* **2022**, *81*, 101735. [CrossRef] [PubMed]
4. Salminen, A.; Kauppinen, A.; Hyttinen, J.M.; Toropainen, E.; Kaarniranta, K. Endoplasmic reticulum stress in age-related macular degeneration: Trigger for neovascularization. *Mol. Med.* **2010**, *16*, 535–542. [CrossRef]
5. Whitmore, S.S.; Sohn, E.H.; Chirco, K.R.; Drack, A.V.; Stone, E.M.; Tucker, B.A.; Mullins, R.F. Complement activation and choriocapillaris loss in early AMD: Implications for pathophysiology and therapy. *Prog. Retin. Eye Res.* **2015**, *45*, 1–29. [CrossRef] [PubMed]
6. Curcio, C.A. Soft Drusen in Age-Related Macular Degeneration: Biology and Targeting via the Oil Spill Strategies. *Investig. Ophthalmol. Vis. Sci.* **2018**, *59*, AMD160–AMD181. [CrossRef]
7. Krenitsky, V.P.; Delgado, M.; Nadolny, L.; Sahasrabudhe, K.; Ayala, L.; Clareen, S.S.; Hilgraf, R.; Albers, R.; Kois, A.; Hughes, K.; et al. Aminopurine based JNK inhibitors for the prevention of ischemia reperfusion injury. *Bioorg. Med. Chem. Lett.* **2012**, *22*, 1427–1432. [CrossRef]
8. Nakano, R.; Nakayama, T.; Sugiya, H. Biological Properties of JNK3 and Its Function in Neurons, Astrocytes, Pancreatic β-Cells and Cardiovascular Cells. *Cells* **2020**, *9*, 1802. [CrossRef]
9. De Los Reyes Corrales, T.; Losada-Pérez, M.; Casas-Tintó, S. JNK Pathway in CNS Pathologies. *Int. J. Mol. Sci.* **2021**, *22*, 3883. [CrossRef]
10. Kim, B.J.; Zack, D.J. The Role of c-Jun N-Terminal Kinase (JNK) in Retinal Degeneration and Vision Loss. *Adv. Exp. Med. Biol.* **2018**, *1074*, 351–357. [CrossRef]
11. Du, H.; Sun, X.; Guma, M.; Luo, J.; Ouyang, H.; Zhang, X.; Zeng, J.; Quach, J.; Nguyen, D.H.; Shaw, P.X.; et al. JNK inhibition reduces apoptosis and neovascularization in a murine model of age-related macular degeneration. *Proc. Natl. Acad. Sci. USA* **2013**, *110*, 2377–2382. [CrossRef] [PubMed]
12. Guma, M.; Rius, J.; Duong-Polk, K.X.; Haddad, G.G.; Lindsey, J.D.; Karin, M. Genetic and pharmacological inhibition of JNK ameliorates hypoxia-induced retinopathy through interference with VEGF expression. *Proc. Natl. Acad. Sci. USA* **2009**, *106*, 8760–8765. [CrossRef] [PubMed]
13. Zhao, Y.; Ding, M.; Yan, F.; Yin, J.; Shi, W.; Yang, N.; Zhao, H.; Fang, Y.; Huang, Y.; Zheng, Y.; et al. Inhibition of the JAK2/STAT3 pathway and cell cycle re-entry contribute to the protective effect of remote ischemic pre-conditioning of rat hindlimbs on cerebral ischemia/reperfusion injury. *CNS Neurosci. Ther.* **2022**, *2022*, 14023. [CrossRef]
14. Carboni, S.; Antonsson, B.; Gaillard, P.; Gotteland, J.P.; Gillon, J.Y.; Vitte, P.A. Control of death receptor and mitochondrial-dependent apoptosis by c-Jun N-terminal kinase in hippocampal CA1 neurones following global transient ischaemia. *J. Neurochem.* **2005**, *92*, 1054–1060. [CrossRef] [PubMed]
15. Gehringer, M.; Muth, F.; Koch, P.; Laufer, S.A. c-Jun N-terminal kinase inhibitors: A patent review (2010–2014). *Expert Opin. Ther. Pat.* **2015**, *25*, 849–872. [CrossRef] [PubMed]
16. Irving, E.A.; Bamford, M. Role of mitogen- and stress-activated kinases in ischemic injury. *J. Cereb. Blood Flow. Metab.* **2002**, *22*, 631–647. [CrossRef] [PubMed]
17. Kuan, C.Y.; Burke, R.E. Targeting the JNK signaling pathway for stroke and Parkinson's diseases therapy. *Curr. Drug Targets CNS Neurol. Disord.* **2005**, *4*, 63–67. [CrossRef]
18. Pan, J.; Zhang, Q.G.; Zhang, G.Y. The neuroprotective effects of K252a through inhibiting MLK3/MKK7/JNK3 signaling pathway on ischemic brain injury in rat hippocampal CA1 region. *Neuroscience* **2005**, *131*, 147–159. [CrossRef]
19. Schepetkin, I.A.; Kirpotina, L.N.; Khlebnikov, A.I.; Hanks, T.S.; Kochetkova, I.; Pascual, D.W.; Jutila, M.A.; Quinn, M.T. Identification and characterization of a novel class of c-Jun N-terminal kinase inhibitors. *Mol. Pharmacol.* **2012**, *81*, 832–845. [CrossRef]
20. Atochin, D.N.; Schepetkin, I.A.; Khlebnikov, A.I.; Seledtsov, V.I.; Swanson, H.; Quinn, M.T.; Huang, P.L. A novel dual NO-donating oxime and c-Jun N-terminal kinase inhibitor protects against cerebral ischemia-reperfusion injury in mice. *Neurosci. Lett.* **2016**, *618*, 45–49. [CrossRef]
21. Carboni, S.; Hiver, A.; Szyndralewiez, C.; Gaillard, P.; Gotteland, J.P.; Vitte, P.A. AS601245 (1,3-benzothiazol-2-yl (2-[[2-(3-pyridinyl) ethyl] amino]-4 pyrimidinyl) acetonitrile): A c-Jun NH_2-terminal protein kinase inhibitor with neuroprotective properties. *J. Pharmacol. Exp. Ther.* **2004**, *310*, 25–32. [CrossRef] [PubMed]

22. Gao, Y.; Signore, A.P.; Yin, W.; Cao, G.; Yin, X.M.; Sun, F.; Luo, Y.; Graham, S.H.; Chen, J. Neuroprotection against focal ischemic brain injury by inhibition of c-Jun N-terminal kinase and attenuation of the mitochondrial apoptosis-signaling pathway. *J. Cereb. Blood Flow Metab.* **2005**, *25*, 694–712. [CrossRef] [PubMed]
23. Guan, Q.H.; Pei, D.S.; Zhang, Q.G.; Hao, Z.B.; Xu, T.L.; Zhang, G.Y. The neuroprotective action of SP600125, a new inhibitor of JNK, on transient brain ischemia/reperfusion-induced neuronal death in rat hippocampal CA1 via nuclear and non-nuclear pathways. *Brain Res.* **2005**, *1035*, 51–59. [CrossRef]
24. Koch, P.; Gehringer, M.; Laufer, S.A. Inhibitors of c-Jun N-terminal kinases: An update. *J. Med. Chem.* **2015**, *58*, 72–95. [CrossRef]
25. Schepetkin, I.A.; Kirpotina, L.N.; Hammaker, D.; Kochetkova, I.; Khlebnikov, A.I.; Lyakhov, S.A.; Firestein, G.S.; Quinn, M.T. Anti-Inflammatory Effects and Joint Protection in Collagen-Induced Arthritis after Treatment with IQ-1S, a Selective c-Jun N-Terminal Kinase Inhibitor. *J. Pharmacol. Exp. Ther.* **2015**, *353*, 505–516. [CrossRef] [PubMed]
26. Schepetkin, I.A.; Khlebnikov, A.I.; Potapov, A.S.; Kovrizhina, A.R.; Matveevskaya, V.V.; Belyanin, M.L.; Atochin, D.N.; Zanoza, S.O.; Gaidarzhy, N.M.; Lyakhov, S.A.; et al. Synthesis, biological evaluation, and molecular modeling of 11*H*-indeno[1,2-*b*]quinoxalin-11-one derivatives and tryptanthrin-6-oxime as c-Jun N-terminal kinase inhibitors. *Eur. J. Med. Chem.* **2019**, *161*, 179–191. [CrossRef]
27. Plotnikov, M.B.; Chernysheva, G.A.; Smolyakova, V.I.; Aliev, O.I.; Trofimova, E.S.; Sherstoboev, E.Y.; Osipenko, A.N.; Khlebnikov, A.I.; Anfinogenova, Y.J.; Schepetkin, I.A.; et al. Neuroprotective Effects of a Novel Inhibitor of c-Jun N-Terminal Kinase in the Rat Model of Transient Focal Cerebral Ischemia. *Cells* **2020**, *9*, 1860. [CrossRef]
28. Plotnikov, M.B.; Chernysheva, G.A.; Aliev, O.I.; Smol'iakova, V.I.; Fomina, T.I.; Osipenko, A.N.; Rydchenko, V.S.; Anfinogenova, Y.J.; Khlebnikov, A.I.; Schepetkin, I.A.; et al. Protective Effects of a New C-Jun N-terminal Kinase Inhibitor in the Model of Global Cerebral Ischemia in Rats. *Molecules* **2019**, *24*, 1722. [CrossRef]
29. Zhdankina, A.A.; Fursova, A.Z.h.; Logvinov, S.V.; Kolosova, N.G. Clinical and morphological characteristics of chorioretinal degeneration in early aging OXYS rats. *Bull. Exp. Biol. Med.* **2008**, *146*, 455–458. [CrossRef]
30. Markovets, A.M.; Saprunova, V.B.; Zhdankina, A.A.; Fursova, A.Z.h.; Bakeeva, L.E.; Kolosova, N.G. Alterations of retinal pigment epithelium cause AMD-like retinopathy in senescence-accelerated OXYS rats. *Aging* **2011**, *3*, 44–54. [CrossRef]
31. Telegina, D.V.; Kozhevnikova, O.S.; Bayborodin, S.I.; Kolosova, N.G. Contributions of age-related alterations of the retinal pigment epithelium and of glia to the AMD-like pathology in OXYS rats. *Sci. Rep.* **2017**, *7*, 41533. [CrossRef] [PubMed]
32. Brecht, S.; Kirchhof, R.; Chromik, A.; Willesen, M.; Nicolaus, T.; Raivich, G.; Wessig, J.; Waetzig, V.; Goetz, M.; Claussen, M.; et al. Specific pathophysiological functions of JNK isoforms in the brain. *Eur. J. Neurosci.* **2005**, *21*, 363–377. [CrossRef] [PubMed]
33. Kolosova, N.G.; Muraleva, N.A.; Zhdankina, A.A.; Stefanova, N.A.; Fursova, A.Z.; Blagosklonny, M.V. Prevention of age-related macular degeneration-like retinopathy by rapamycin in rats. *Am. J. Pathol.* **2012**, *181*, 472–477. [CrossRef] [PubMed]
34. Kolosova, N.G.; Stefanova, N.A.; Korbolina, E.E.; Fursova, A.Z.h.; Kozhevnikova, O.S. The senescence-accelerated oxys rats—A genetic model of premature aging and age-dependent degenerative diseases. *Adv. Gerontol.* **2014**, *27*, 336–340. [PubMed]
35. Taylor, C.R.; Rudbeck, L.; Corporation, D. *Immunohistochemical Staining Methods*, 6th ed.; DAKO Corporation: Glostrup, Denmark, 2013.
36. Markovets, A.M.; Fursova, A.Z.; Kolosova, N.G. Therapeutic action of the mitochondria-targeted antioxidant SkQ1 on retinopathy in OXYS rats linked with improvement of VEGF and PEDF gene expression. *PLoS ONE* **2011**, *6*, e21682. [CrossRef]
37. Kozhevnikova, O.S.; Korbolina, E.E.; Ershov, N.I.; Kolosova, N.G. Rat retinal transcriptome: Effects of aging and AMD-like retinopathy. *Cell Cycle* **2013**, *12*, 1745–1761. [CrossRef]
38. Gelfand, B.D.; Ambati, J. A Revised Hemodynamic Theory of Age-Related Macular Degeneration. *Trends Mol. Med.* **2016**, *22*, 656–670. [CrossRef]
39. Lipecz, A.; Miller, L.; Kovacs, I.; Czakó, C.; Csipo, T.; Baffi, J.; Csiszar, A.; Tarantini, S.; Ungvari, Z.; Yabluchanskiy, A.; et al. Microvascular contributions to age-related macular degeneration (AMD): From mechanisms of choriocapillaris aging to novel interventions. *Geroscience* **2019**, *41*, 813–845. [CrossRef]
40. Zarbin, M.A. Current concepts in the pathogenesis of age-related macular degeneration. *Arch. Ophthalmol.* **2004**, *122*, 598–614. [CrossRef]
41. Kozhevnikova, O.S.; Fursova, A.Z.; Markovets, A.M.; Telegina, D.V.; Muraleva, N.A.; Kolosova, N.G. VEGF and PEDF levels in the rat retina: Effects of aging and AMD-like retinopathy. *Adv. Gerontol.* **2018**, *31*, 339–344.
42. Polato, F.; Becerra, S.P. Pigment Epithelium-Derived Factor, a Protective Factor for Photoreceptors in Vivo. *Adv. Exp. Med. Biol.* **2016**, *854*, 699–706. [CrossRef] [PubMed]
43. Hu, M.C.; Qiu, W.R.; Wang, Y.P. JNK1, JNK2 and JNK3 are p53 N-terminal serine 34 kinases. *Oncogene* **1997**, *15*, 2277–2287. [CrossRef]
44. Kaarniranta, K.; Uusitalo, H.; Blasiak, J.; Felszeghy, S.; Kannan, R.; Kauppinen, A.; Salminen, A.; Sinha, D.; Ferrington, D. Mechanisms of mitochondrial dysfunction and their impact on age-related macular degeneration. *Prog. Retin. Eye Res.* **2020**, *79*, 100858. [CrossRef] [PubMed]
45. Kolosova, N.G.; Aidagulova, S.V.; Nepomnyashchikh, G.I.; Shabalina, I.G.; Shalbueva, N.I. Dynamics of structural and functional changes in hepatocyte mitochondria of senescence-accelerated OXYS rats. *Bull. Exp. Biol. Med.* **2001**, *132*, 814–819. [CrossRef] [PubMed]
46. Vays, V.B.; Eldarov, C.M.; Vangely, I.M.; Kolosova, N.G.; Bakeeva, L.E.; Skulachev, V.P. Antioxidant SkQ1 delays sarcopenia-associated damage of mitochondrial ultrastructure. *Aging* **2014**, *6*, 140–148. [CrossRef]

47. El'darov, C.M.; Vays, V.B.; Vangeli, I.M.; Kolosova, N.G.; Bakeeva, L.E. Morphometric Examination of Mitochondrial Ultrastructure in Aging Cardiomyocytes. *Biochemistry* **2015**, *80*, 604–609. [CrossRef]
48. Tyumentsev, M.A.; Stefanova, N.A.; Kiseleva, E.V.; Kolosova, N.G. Mitochondria with Morphology Characteristic for Alzheimer's Disease Patients Are Found in the Brain of OXYS Rats. *Biochemistry* **2018**, *83*, 1083–1088. [CrossRef]
49. Shore, G.C.; Papa, F.R.; Oakes, S.A. Signaling cell death from the endoplasmic reticulum stress response. *Curr. Opin. Cell Biol.* **2011**, *23*, 143–149. [CrossRef]
50. Davis, R.J. Signal transduction by the JNK group of MAP kinases. *Cell* **2000**, *103*, 239–252. [CrossRef]

biomedicines

MDPI

Article

Association between Age-Related Macular Degeneration and the Risk of Diabetes Mellitus: A Nationwide Cohort Study

Wonyoung Jung [1,2], Je Moon Yoon [3,*], Kyungdo Han [4], Bongseong Kim [4], Sungsoon Hwang [3,5], Dong Hui Lim [3,5] and Dong Wook Shin [1,5,*]

1 Department of Family Medicine and Supportive Care Center, Samsung Medical Center, Sungkyunkwan University School of Medicine, Seoul 06351, Korea
2 Department of Medicine, Sungkyunkwan University School of Medicine, Seoul 06351, Korea
3 Department of Ophthalmology, Samsung Medical Center, Sungkyunkwan University School of Medicine, Seoul 06351, Korea
4 Department of Statistics and Actuarial Science, Soongsil University, Seoul 06978, Korea
5 Department of Clinical Research Design & Evaluation, Samsung Advanced Institute for Health Science & Technology (SAIHST), Sungkyunkwan University, Seoul 06355, Korea
* Correspondence: jemun1010@gmail.com (J.M.Y.); dwshin.md@gmail.com (D.W.S.); Tel.: +82-2-3410-3563 (J.M.Y.); +82-2-3410-5252 (D.W.S.); Fax: +82-2-3410-0074 (J.M.Y.); +82-2-3410-0388 (D.W.S.)

Abstract: Age-related macular degeneration (AMD) is a degenerative and progressive disease of the macula, the part of the retina that is responsible for central vision. AMD shares some risk factors with diabetes mellitus (DM), but little is known about the risk of DM in individuals with AMD. With the goal of establishing novel perspectives, this study aimed to investigate the association between AMD and the risk of DM using the Korean Nationwide Health Insurance Database. Individuals aged ≥ 50 years who underwent a national health screening program in 2009 were enrolled. Participants were categorized by the presence of AMD and visual disability (VD). The Cox hazard regression model was used to examine hazard ratios (HRs) of DM with adjustment for potential confounders. Stratified analyses by age, sex, and comorbidities (hypertension or dyslipidemia) were also performed. During a mean follow-up of 8.61 years, there were 403,367 (11.76%) DM incidences among the final 3,430,532 participants. The crude HR (95% confidence interval (CI)) was 1.16 (1.13–1.20) for AMD. After adjusting for potential confounders, AMD was associated with a 3% decreased risk of DM (aHR 0.97, 95% CI 0.95–1.00), but no significant association with the risk of DM was found in AMD with VD (aHR 1.03, 95% CI 0.93–1.14). In summary, we did not find an increased risk of DM in individuals with AMD. A 3% decreased risk of DM in patients with AMD is not clinically meaningful. Our study suggests that the association between AMD and the risk of DM is weak, considering the potential confounders. Further studies examining this association are needed to extend our knowledge.

Keywords: age-related macular degeneration; diabetes mellitus; visual disability

Citation: Jung, W.; Yoon, J.M.; Han, K.; Kim, B.; Hwang, S.; Lim, D.H.; Shin, D.W. Association between Age-Related Macular Degeneration and the Risk of Diabetes Mellitus: A Nationwide Cohort Study. *Biomedicines* **2022**, *10*, 2435. https://doi.org/10.3390/biomedicines10102435

Academic Editor: Oyuna Kozhevnikova

Received: 28 August 2022
Accepted: 26 September 2022
Published: 29 September 2022

1. Introduction

Age-related macular degeneration (AMD) is a degenerative and progressive disease of the macula, the part of the retina that is responsible for central vision [1]. Since the number of people with AMD worldwide is expected to increase to 288 million by 2040 [2], it is important to identify comorbid conditions and health problems among those affected.

AMD is associated with traditional risk factors for cardiovascular disease, such as aging, cigarette smoking, obesity, elevated levels of cholesterol, and hypertension [3–9], each of which also is a risk factor for diabetes mellitus (DM). Previous studies have focused on the risk of AMD in individuals with DM [10–13], suggesting a link through chronic inflammation and oxidative stress. However, little is known about the risk of DM in people with AMD.

Notably, AMD is the leading cause of blindness in developed countries, particularly in patients over 60 years of age [2]. Visual disability (VD) is associated with a broad range of chronic conditions, including cardiometabolic, neuropsychiatric, and musculoskeletal disorders [14,15]. A nearly twofold risk of type 2 DM among people aged 65 or older with visual impairment was reported in a US claims database study [14]. In addition, a recent prospective study highlighted the role of the neighborhood environment in DM risk in people with visual impairment [16]. Environmental barriers in accessing healthcare service, proper nutrition, and recreational resources that promote positive lifestyle behaviors may contribute to the increased risk of type 2 DM in people with visual impairment. These findings suggest an association between AMD, the leading cause of visual disability, and the risk of DM. With the goal of establishing novel perspectives, we aimed to investigate the association between AMD, which is further categorized by VD status, and the risk of DM incidence using the Korean nationwide claims database.

2. Materials and Methods

2.1. Data Source and Study Setting

The National Health Insurance (NHI) in Korea is a mandatory social health insurance program and provides universal coverage to 97% of the Korean population. Due to their low-income status, the remaining 3% are covered by Medicaid, which is funded by the general Korean tax. The National Health Insurance Service (NHIS) is a public corporation that administers to both NHI enrollees and Medicaid beneficiaries. For every individual aged 40 or above, the NHIS provides a biennial national health screening that consists of a self-questionnaire on health behavior (smoking, drinking, and past medical history), anthropometric measurements (blood pressure, body mass index), and laboratory test findings (fasting glucose, serum lipid levels) [17]. Therefore, the NHIS database is an eligibility database (age, sex, disability type and severity, socioeconomic variables, income level, and type of eligibility), a medical treatment database (based on medical bills claimed by medical service providers for medical expenses), and a health screening database (results of national health screening). The NHIS database has been widely used in epidemiologic studies in Korea [18].

2.2. Study Population

A total of 4,470,729 participants aged $\geq$50 years underwent general health screenings in 2009. To extract a homogeneous group of individuals devoid of DM or any glucose metabolism disorders, we excluded participants with DM with the following criteria: (1) history of antidiabetic medication prescription (metformin, sulfonylurea, thiazolidinedione, sodium-glucose cotransporter-2 (SGLT-2) inhibitor, dipeptidyl peptidase-4 (DPP-4) inhibitor, alpha-glucosidase, and/or insulin) before health screening in 2009 (n = 576,759); (2) history of type 1 DM (n = 37,877) or gestational DM (n = 58); and (3) serum fasting glucose level of 126 mg/dL or above (n = 188,875) at health screening in 2009. We also excluded patients who were diagnosed with DM (n = 17,392) or died (n = 13,628) within 1 year from health screening in 2009. Participants with missing data on at least one variable used in the analysis (n = 205,608) were also excluded. As a result, a total of 3,430,532 participants were included in the final analyses.

2.3. Definition of Age-Related Macular Degeneration

Individuals with AMD were identified based on the International Classification of Diseases, 10th Revision (ICD-10), code for AMD (H353) by an ophthalmologist within 1 year before the health screening examination. This operational definition of AMD has been used in previous epidemiologic studies on AMD [19,20].

2.4. Definition of Visual Disability

According to the National Disability Registration System (NDRS) in Korea, VD refers to a visual loss or visual field defect. Registration for disability requires submission of validated documentation of the results of disability diagnosis by a specialist physician [21]. For instance, a best-corrected visual acuity (BCVA) of 0.02 or less using the Snellen visual acuity chart is a minimum requirement to register for VD. In Korea, the severity of disability is graded from 1 (most severe) to 6 (least severe) according to predefined criteria (Supplementary Table S1), and welfare benefits, including disability pension, are determined based on that severity score [22]. Therefore, almost all individuals with disability apply for registration and are included in the National Disability Registration System in Korea.

2.5. Study Outcomes and Follow-Up

The study endpoint was incidence of DM, followed by the period of time analyzed. Newly diagnosed DM was defined as a prescription of antidiabetic medication that was linked to a DM ICD-10 code (E11–E14). The participants were followed from the date of health screening examination in 2009 to the date of DM incidence or death or until the end of the study period (31 December 2019), whichever came first.

2.6. Covariates

During the health screening, the participants provided information on lifestyle behaviors using standardized questionnaires [17]. Smoking status was categorized as non-, ex-, and current smoker. Alcohol consumption was categorized as none, mild, and heavy. Heavy drinking was defined as $\geq$30 g of alcohol consumption per day. Individuals were considered to have regular physical activity if they exercised strenuously $\geq$ 1 time/week for at least 20 min per session. Household income was dichotomized by the lowest 20 percentile according to the health insurance premium determined by income status and not by health status in the social health insurance system in Korea. Body mass index (BMI) was calculated as weight in kilograms divided by height in meters squared.

Fasting serum glucose level, serum hemoglobin level, and estimated glomerular filtration rate (eGFR) were also assessed. Hypertension was defined as any of the following: systolic blood pressure $\geq$ 140 mmHg, diastolic blood pressure $\geq$ 90 mmHg, or treatment with an antihypertensive medication that was linked to hypertension ICD-10 codes (I10–I13 and I15), and resulted in at least one claim in 1 year. Dyslipidemia was defined as total cholesterol $\geq$ 240 mg/dL or a history of a lipid-lowering medication that was associated with an ICD-10 code (E78). Finally, the Charlson Comorbidity Index (CCI) was calculated based on the diagnosis code [23].

2.7. Statistical Analysis

The comparison of baseline characteristics by the presence of AMD and VD was conducted using Student's t-test for continuous variables or the chi-square test for categorical variables. The incidence rates of DM were calculated as incident cases divided by 100,000 person-years. The Cox hazard regression model was used to examine the hazard ratios (HRs) of DM. Multivariable analyses were serially adjusted for age and sex (Model 2); household income, area of residence, BMI, smoking, alcohol consumption, and regular exercise (Model 3); and fasting serum glucose level, serum hemoglobin level, eGFR, hypertension, dyslipidemia, and CCI (Model 4). These variables are based on the method used in previous studies [10,24–26]. P for trend was calculated among the HRs of control, AMD without VD, and AMD with VD groups. Finally, to evaluate the potential effects of modification by age, sex, and comorbidity status, P for interaction was calculated using stratified analysis.

Statistical analyses were performed using SAS version 9.4 (SAS Institute Inc., Cary, NC, USA). A p-value < 0.05 was considered statistically significant.

2.8. Ethics Statement

This study was approved by the Institutional Review Board of the Samsung Medical Center (SMC 2022-03-060). Anonymized and de-identified information was used for analyses; therefore, informed consent was not required. The database is open to all researchers whose study protocols are approved by the official review committee.

3. Results

3.1. Baseline Characteristics

A total of 3,430,532 participants were enrolled in the final analysis. At baseline, 38,726 (1.13%) participants had AMD, and 2723 (7.03% of AMD) had visual disability (Table 1). The AMD group was older with larger proportions of women and nonsmokers and had higher prevalence of hypertension and dyslipidemia compared with the non-AMD group (all $p < 0.001$). In addition, the AMD group had lower alcohol consumption, total cholesterol level, and lower income and resided less frequently in urban areas than the non-AMD group (all $p < 0.001$).

Table 1. Baseline characteristics of the study population according to the presence of age-related macular degeneration.

	Age-Related Macular Degeneration			Age-Related Macular Degeneration		
	Absent (*n* = 3,391,806)	Present (*n* = 38,726)	*p*-Value	Without VD (*n* = 36,003)	With VD (*n* = 2723)	*p*-Value
Mean age, years	60.09 ± 8.19	67.06 ± 8.56	<0.001	67.02 ± 8.57	67.61 ± 8.50	<0.001
Sex, male	1,588,565 (46.8)	15,889 (41.0)	<0.001	14,623 (40.6)	1266 (46.5)	<0.001
Smoking			<0.001			<0.001
Nonsmoker	2,294,086 (67.6)	28,673 (74.0)		26,724 (74.2)	1949 (71.6)	
Ex-smoker	519,477 (15.3)	5950 (15.4)		5515 (15.3)	435 (16.0)	
Current smoker	578,243 (17.1)	4103 (10.6)		3764 (10.5)	339 (12.5)	
Alcohol consumption			<0.001			<0.001
None	2,208,593 (65.1)	29,238 (75.5)		27,174 (75.5)	2064 (75.8)	
Mild	978,072 (28.8)	8041 (20.8)		7495 (20.8)	546 (20.1)	
Heavy	205,141 (6.1)	1447 (3.7)		1334 (3.7)	113 (4.2)	
Regular physical activity	709,047 (20.9)	7861 (20.3)	0.004	7295 (20.3)	566 (20.8)	0.012
Anthropometrics						
Body mass index, kg/m^2	23.94 ± 2.95	23.85 ± 2.99	<0.001	23.85 ± 2.99	23.80 ± 3.01	<0.001
WC, cm	81.57 ± 8.19	82.05 ± 8.23	<0.001	82.02 ± 8.22	82.46 ± 8.29	<0.001
Systolic BP, mmHg	125.82 ± 15.68	127.55 ± 15.68	<0.001	127.58 ± 15.66	127.24 ± 15.88	<0.001
Diastolic BP, mmHg	77.76 ± 10.14	77.54 ± 9.94	<0.001	77.55 ± 9.93	77.48 ± 10.07	0.052
Comorbidity						
Hypertension	1,414,845 (41.7)	21,292 (55.0)	<0.001	19,851 (55.1)	1441 (52.9)	<0.001
Dyslipidemia	857,795 (25.3)	11,906 (30.7)	<0.001	11,100 (30.8)	806 (29.6)	<0.001
Laboratory findings						
Glucose, fasting, mg/dL	94.63 ± 11.55	94.58 ± 11.47	0.463	94.58 ± 11.48	94.68 ± 11.36	0.628
eGFR, mL/min/1.73 m^2	83.13 ± 33.27	80.71 ± 35.12	<0.001	80.70 ± 35.16	80.78 ± 34.53	<0.001
Total cholesterol, mg/dL	202.30 ± 37.37	201.40 ± 37.94	<0.001	201.47 ± 37.94	200.39 ± 37.91	<0.001
Triglycerides [a], mg/dL	117.49 (117.43–117.56)	116.97 (116.38–117.55)	0.227	116.81 (116.21–117.42)	119.03 (116.80–121.31)	0.049
HDL-C, mg/dL	56.13 ± 31.35	55.95 ± 33.58	0.072	56.09 ± 33.92	54.13 ± 28.62	0.004
LDL-C, mg/dL	120.70 ± 38.57	120.79 ± 38.57	0.665	120.79 ± 38.37	120.75 ± 41.21	0.898
Hemoglobin, mg/dL	13.67 ± 1.45	13.41 ± 1.42	<0.001	13.40 ± 1.42	13.43 ± 1.41	<0.001
Urban residency	1,545,555 (45.6)	15,781 (40.8)	<0.001	14,664 (40.7)	1117 (41.0)	<0.001
Income of lowest 20%	727,814 (21.5)	6931 (17.9)	<0.001	6407 (17.8)	524 (19.2)	<0.001
CCI	1.00 ± 1.15	1.47 ± 1.35	<0.001	1.47 ± 1.35	1.49 ± 1.33	<0.001

Data are expressed as the mean ± standard deviation or number (%). VD, visual disability; WC, waist circumference; BP, blood pressure; eGFR, estimated glomerular filtration rate; HDL-C, high-density lipoprotein cholesterol; LDL-C, low-density lipoprotein cholesterol; CCI, Charlson Comorbidity Index. [a] Geometric mean (95% confidence interval).

The AMD with VD group was older with larger proportions of men and current smokers and lower prevalence of hypertension and dyslipidemia compared with the AMD without VD group (all $p < 0.001$).

3.2. Risk of DM by AMD

During the mean 8.61 years of follow-up, there were 403,367 new cases of DM (11.76%). After considering the covariates, AMD was associated with a 3% decreased risk of DM incidence (aHR 0.95, 95% CI (0.95–1.00), p = 0.041; Table 2). Further analysis that specified AMD by VD status showed that AMD without VD was associated with a 3% lower risk for DM (aHR 0.97, 95% CI (0.94–1.00)), and no significant association with DM was found in AMD with VD (aHR 1.03, 95% CI (0.93–1.14)).

Table 2. Association between age-related macular degeneration and the risk of diabetes mellitus according to visual disability status.

	Subjects (N)	Case (*n*)	IR Per 1000 Person-Years	Model 1 (Crude) HR (95% CI)	Model 2 aHR (95% CI)	Model 3 aHR (95% CI)	Model 4 aHR (95% CI)
Non-AMD	3,391,806	398,308	13.6	1 (Ref.)	1 (Ref.)	1 (Ref.)	1 (Ref.)
AMD	38,726	5059	15.8	1.16 (1.13–1.20)	1.00 (0.97–1.03)	1.01 (0.98–1.04)	0.97 (0.95–1.00)
p-Value				<0.001	0.939	0.420	0.041
AMD without VD	36,003	4684	15.7	1.16 (1.12–1.19)	1.00 (0.97–1.03)	1.01 (0.98–1.04)	0.97 (0.94–1.00)
AMD with VD	2723	375	16.9	1.25 (1.13–1.38)	1.06 (0.95–1.17)	1.06 (0.96–1.17)	1.03 (0.93–1.14)
† *p* for trend				<0.001	0.756	0.312	0.087

IR, incidence rate; PY, person-years; HR, hazard ratio; aHR, adjusted hazard ratio; CI, confidence interval. † *p* for trend was calculated by the (adjusted) hazard ratios among the control, AMD without VD, and AMD with VD groups. Model 2 was adjusted for age and sex. Model 3 was adjusted for age, sex, income, area of residence, body mass index, smoking, alcohol consumption, and regular exercise. Model 4 was adjusted for age, sex, income, area of residence, body mass index, smoking, alcohol consumption, regular exercise, fasting serum glucose level, serum hemoglobin level, eGFR, hypertension, dyslipidemia, and CCI.

3.3. Stratified Analyses

No significant interactions were found in stratified analyses according to age, sex, and comorbidities (hypertension or dyslipidemia) (Figure 1).

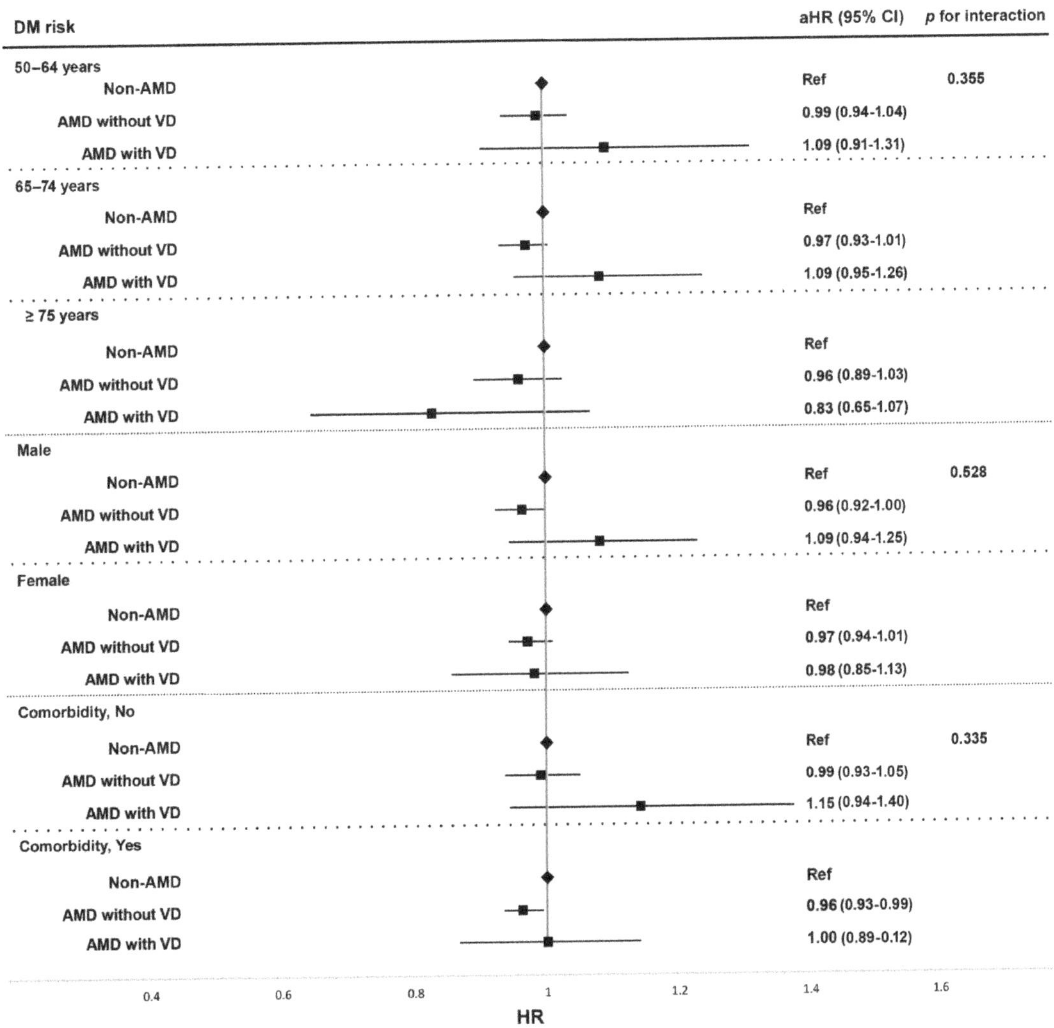

Figure 1. Forest plot of analyses stratified by age, sex, and comorbidity. HRs were adjusted for age, sex, income, area of residence, BMI, smoking, alcohol consumption, regular exercise, hypertension, dyslipidemia, CCI, fasting glucose level, serum hemoglobin level, and eGFR. Non-AMD group was reference (Ref). BMI, body mass index; CCI, Charlson Comorbidity Index; eGFR, estimated glomerular filtration rate; HR, hazard ratio; CI, confidence interval.

4. Discussion

Here, we investigated the association between AMD and the risk of DM. The inclusion of more than three million participants from the NHIS database enabled us to adjust for various confounders. Therefore, we had enough data to detect significant differences, even with stratified analyses. Our study revealed that AMD was associated with a 3% decreased risk of DM, which is not clinically important. No significant effect modifiers were found. Therefore, the association between AMD and the risk of DM is weak.

At baseline, the differences in demographic, socioeconomic, and lifestyle behavior variables between AMD and non-AMD group were noted. Old age, the strongest risk factor of AMD [1], and higher prevalence of hypertension, one of the moderate and consistent risk factors of AMD [1], were associated with AMD in univariate analysis. Large proportions of nonsmokers and nondrinkers and less frequent urban residency in the AMD group may be attributed to older age in the AMD group (67.06 ± 8.56 years) than the non-AMD group (60.09 ± 8.19 years). For total cholesterol level, although the difference was significant ($p < 0.001$) due to the large sample size, it was not clinically important. The smaller proportion of those in the lower 20% income in the AMD without VD group than the AMD with VD group can be explained by the limited access to healthcare service.

There are several pathophysiological mechanisms that suggest a possible link between AMD and DM. Chronic inflammation and oxidative stress may explain the link between AMD and the risk of DM. Oxidative stress triggers outer blood retinal barrier degeneration that leads to AMD [27,28], and oxidative stress is a causative factor in the development of insulin resistance [29,30]. In vitro overstimulation of interleukin 17 receptor C (IL17RC), of which a high level is detected on the surface of peripheral blood cells from patients with AMD, was shown to cause PI3K/Akt/GSK3 insensitivity, high GSK3 activity associated with insulin resistance, and type 2 DM [31]. A transcriptome-wide association study found that the BCAR1, CFDP1, and TMEM170A genes overlap with significant genome-wide association study signals of type 2 DM [32], supporting the potential link between AMD and DM. In addition, the two diseases share common risk factors, such as aging, obesity, and unhealthy lifestyle [33,34].

To our knowledge, this is the first study to report DM risk in people with AMD. Several epidemiologic studies have focused on the risk of AMD in DM patients, and the results of those studies are inconsistent [10–13,26,35,36]: higher risk of neovascular AMD [11] or late AMD [12] was reported in some studies, whereas no increased risk of AMD [13,35] was reported in another. Here, our study suggests a slightly lower risk of DM in AMD patients, contrary to our expectations. The reason for this negative association is not clear; however, one possible explanation is that about 17% of participants with a previous history of DM were excluded from our study population. Excluding participants with a DM diagnosis before baseline decreased the likelihood of individuals with AMD who were 50 years of age or older developing DM, as participants with a high likelihood of developing DM were eliminated before analysis. A report from using the Korean NHIS database (2003–2012) showed that the average age at diagnosis of DM was 64 ± 15 years [37]. According to the Korean Retina Society, the average age at diagnosis of AMD was 69.7 ± 8.0 years [38].

Since the association between AMD and DM was nullified in our final Model 4 in which covariates were serially adjusted from Model 1 (crude), other confounders may explain the association. Therefore, shared risk factors among AMD and DM may explain the association between AMD and DM rather than a common pathophysiology.

Another possible explanation is that lifestyle modification (smoking cessation, physical activity, diet, and nutritional support) prescribed in individuals with AMD may have preventive effects on the development of DM [39]. However, these observations do not fully explain the null or even negative association, and further studies are needed to replicate our finding and to investigate pathophysiological mechanisms.

It is also noteworthy that AMD patients with VD have a slightly higher risk of DM than AMD patients without VD (aHR 1.03, 95% CI (0.93–1.14); aHR 0.97, 95% CI (0.94–1.00), respectively). The strength of our study includes the use of the NDRS in our analysis. VD is associated with an increased risk of DM [14,15], regardless of AMD pathogenesis. Visual impairment is associated with physical inactivity [40–42], which may have adverse consequences for people with VD. Further functional consequences, including social isolation, daily activity restriction, poor quality of life, and frailty, can also result from visual impairment [43–50], especially in the elderly population. Limited access to information and healthcare facilities in people with visual impairment [51] may also contribute to this association. In addition, visual impairment significantly affects nutritional status, with a higher prevalence of obesity [52]. Further studies are needed to explain these findings.

Several limitations should be considered in interpreting our results. First, AMD might be underdiagnosed, especially during the early stages. Most AMD patients in Korea are diagnosed early without symptoms that can be identified from the claims database. Since we used the operational definition based on an ICD-10 code and not retinal photographs, the association between AMD and DM might be weaker in our findings than the actual association. Therefore, unmeasured confounders could be possible. Second, our results cannot be extended to other ethnic groups, considering the ethnic differences in genetic factors related to AMD [53,54]. Finally, other possible confounders, such as nutrition and dietary factors [55,56], were not controlled.

In summary, we did not find an increased risk of DM in individuals with AMD. Our study suggests that the association between AMD and the risk of DM is weak, considering the potential confounders. Further studies examining this association are needed to extend our knowledge.

Supplementary Materials: The following supporting information can be downloaded at: https://www.mdpi.com/article/10.3390/biomedicines10102435/s1, Table S1: Visual disability criteria of the Republic of Korea.

Author Contributions: Conceptualization, W.J., J.M.Y., K.H. and D.W.S.; Methodology, W.J., J.M.Y., K.H., B.K., S.H., D.H.L. and D.W.S.; Software, K.H. and B.K.; Validation, K.H., B.K. and D.W.S.; Formal analysis, K.H., B.K. and D.W.S.; Investigation, W.J., J.M.Y., B.K. and D.W.S.; Resources, K.H. and B.K.; Data curation, K.H. and B.K.; Writing—original draft preparation, W.J.; Writing—review and editing, W.J., J.M.Y., S.H., D.H.L. and D.W.S.; Visualization, W.J.; Supervision, J.M.Y. and D.W.S.; Project administration, D.H.L. and D.W.S.; Funding acquisition, D.H.L. and D.W.S. All authors have read and agreed to the published version of the manuscript.

Funding: This research was partially supported by a grant from the Korea Health Technology R&D Project through the Korea Health Industry Development Institute (KHIDI), funded by the Ministry of Health and Welfare, Republic of Korea (grant number: HI20C1073), which was received by Dong Wook Shin, and by a grant from the National Research Foundation of Korea, funded by the Ministry of Education, Republic of Korea (NRF-2021R1C1C1007795), which was received by Dong Hui Lim.

Institutional Review Board Statement: This study's protocol was approved by the Institutional Review Board of the Samsung Medical Center (SMC 2022-03-060). Informed consent was waived, becaused anonymized and deidentified data were used.

Informed Consent Statement: Not required due to the study design. Informed consent from individual participants was waived, as this study involved only deidentified data.

Data Availability Statement: The datasets used and/or analyzed during this study are available from the corresponding author on reasonable request.

Conflicts of Interest: The authors declare no conflict of interest.

References

1. Keenan, T.D.L.; Cukras, C.A.; Chew, E.Y. Age-Related Macular Degeneration: Epidemiology and Clinical Aspects. *Adv. Exp. Med. Biol.* **2021**, *1256*, 1–31.
2. Wong, W.L.; Su, X.; Li, X.; Cheung, C.M.; Klein, R.; Cheng, C.Y.; Wong, T.Y. Global prevalence of age-related macular degeneration and disease burden projection for 2020 and 2040: A systematic review and meta-analysis. *Lancet Glob. Health* **2014**, *2*, e106–e116. [CrossRef]
3. Snow, K.K.; Seddon, J.M. Do age-related macular degeneration and cardiovascular disease share common antecedents? *Ophthalmic Epidemiol.* **1999**, *6*, 125–143. [CrossRef]
4. Tan, J.S.; Mitchell, P.; Smith, W.; Wang, J.J. Cardiovascular risk factors and the long-term incidence of age-related macular degeneration: The Blue Mountains Eye Study. *Ophthalmology* **2007**, *114*, 1143–1150. [CrossRef]
5. Delcourt, C.; Michel, F.; Colvez, A.; Lacroux, A.; Delage, M.; Vernet, M.H. Associations of cardiovascular disease and its risk factors with age-related macular degeneration: The POLA study. *Ophthalmic Epidemiol.* **2001**, *8*, 237–249. [CrossRef]
6. Cackett, P.; Wong, T.Y.; Aung, T.; Saw, S.M.; Tay, W.T.; Rochtchina, E.; Mitchell, P.; Wang, J.J. Smoking, cardiovascular risk factors, and age-related macular degeneration in Asians: The Singapore Malay Eye Study. *Am. J. Ophthalmol.* **2008**, *146*, 960–967.e961. [CrossRef]
7. Lüdtke, L.; Jürgens, C.; Ittermann, T.; Völzke, H.; Tost, F. Age-Related Macular Degeneration and Associated Risk Factors in the Population-Based Study of Health in Pomerania (SHIP-Trend). *Med. Sci. Monit.* **2019**, *25*, 6383–6390. [CrossRef]

8. Cheung, C.M.G.; Gan, A.; Fan, Q.; Chee, M.L.; Apte, R.S.; Khor, C.C.; Yeo, I.; Mathur, R.; Cheng, C.Y.; Wong, T.Y.; et al. Plasma lipoprotein subfraction concentrations are associated with lipid metabolism and age-related macular degeneration. *J. Lipid Res.* **2017**, *58*, 1785–1796. [CrossRef]
9. Erke, M.G.; Bertelsen, G.; Peto, T.; Sjølie, A.K.; Lindekleiv, H.; Njølstad, I. Cardiovascular risk factors associated with age-related macular degeneration: The Tromsø Study. *Acta Ophthalmol.* **2014**, *92*, 662–669. [CrossRef]
10. Choi, J.K.; Lym, Y.L.; Moon, J.W.; Shin, H.J.; Cho, B. Diabetes mellitus and early age-related macular degeneration. *Arch. Ophthalmol.* **2011**, *129*, 196–199. [CrossRef]
11. Clemons, T.E.; Milton, R.C.; Klein, R.; Seddon, J.M.; Ferris, F.L., 3rd. Risk factors for the incidence of Advanced Age-Related Macular Degeneration in the Age-Related Eye Disease Study (AREDS) AREDS report no. 19. *Ophthalmology* **2005**, *112*, 533–539.
12. Leske, M.C.; Wu, S.Y.; Hennis, A.; Nemesure, B.; Yang, L.; Hyman, L.; Schachat, A.P. Nine-year incidence of age-related macular degeneration in the Barbados Eye Studies. *Ophthalmology* **2006**, *113*, 29–35. [CrossRef]
13. Voutilainen-Kaunisto, R.M.; Teräsvirta, M.E.; Uusitupa, M.I.; Niskanen, L.K. Age-related macular degeneration in newly diagnosed type 2 diabetic patients and control subjects: A 10-year follow-up on evolution, risk factors, and prognostic significance. *Diabetes Care* **2000**, *23*, 1672–1678. [CrossRef]
14. Kolli, A.; Seiler, K.; Kamdar, N.; De Lott, L.B.; Peterson, M.D.; Meade, M.A.; Ehrlich, J.R. Longitudinal Associations Between Vision Impairment and the Incidence of Neuropsychiatric, Musculoskeletal, and Cardiometabolic Chronic Diseases. *Am. J. Ophthalmol.* **2022**, *235*, 163–171. [CrossRef]
15. Crews, J.E.; Chou, C.F.; Sekar, S.; Saaddine, J.B. The Prevalence of Chronic Conditions and Poor Health Among People With and Without Vision Impairment, Aged $\geq$ 65 Years, 2010–2014. *Am. J. Ophthalmol.* **2017**, *182*, 18–30. [CrossRef]
16. Clarke, P.; Khan, A.M.; Kamdar, N.; Seiler, K.; Latham-Mintus, K.; Peterson, M.D.; Meade, M.A.; Ehrlich, J.R. Risk of Type 2 Diabetes Mellitus among Adults Aging with Vision Impairment: The Role of the Neighborhood Environment. *Disabil. Health J.* **2022**. [CrossRef]
17. Shin, D.W.; Cho, J.; Park, J.H.; Cho, B. National General Health Screening Program in Korea: History, current status, and future direction. *Precis Future Med.* **2022**, *6*, 9–31. [CrossRef]
18. Lee, Y.H.; Han, K.; Ko, S.H.; Ko, K.S.; Lee, K.U. Data Analytic Process of a Nationwide Population-Based Study Using National Health Information Database Established by National Health Insurance Service. *Diabetes Metab. J.* **2016**, *40*, 79–82. [CrossRef]
19. Tsai, D.C.; Chen, S.J.; Huang, C.C.; Yuan, M.K.; Leu, H.B. Age-Related Macular Degeneration and Risk of Degenerative Dementia among the Elderly in Taiwan: A Population-Based Cohort Study. *Ophthalmology* **2015**, *122*, 2327–2335.e2322. [CrossRef]
20. Choi, S.; Jahng, W.J.; Park, S.M.; Jee, D. Association of Age-Related Macular Degeneration on Alzheimer or Parkinson Disease: A Retrospective Cohort Study. *Am. J. Ophthalmol.* **2020**, *210*, 41–47. [CrossRef]
21. Shin, D.W.; Lee, J.W.; Jung, J.H.; Han, K.; Kim, S.Y.; Choi, K.S.; Park, J.H.; Park, J.H. Disparities in Cervical Cancer Screening Among Women With Disabilities: A National Database Study in South Korea. *J. Clin. Oncol.* **2018**, *36*, 2778–2786. [CrossRef]
22. Number of Registered Persons with Disabilities and Disability Pension Recipients, Ministry of Health and Welfare, Republic of Korea, 2022. Available online: https://www.mohw.go.kr/eng/hs/hs0106.jsp?PAR_MENU_ID=1006&MENU_ID=100606 (accessed on 5 April 2022).
23. Kim, K.H. Comparative Study on Three Algorithms of the ICD-10 Charlson Comorbidity Index with Myocardial Infarction Patients. *J. Prev. Med. Public Health* **2010**, *43*, 42–49. [CrossRef]
24. Hahn, P.; Acquah, K.; Cousins, S.W.; Lee, P.P.; Sloan, F.A. Ten-year incidence of age-related macular degeneration according to diabetic retinopathy classification among medicare beneficiaries. *Retina* **2013**, *33*, 911–919. [CrossRef]
25. Blitzer, A.L.; Ham, S.A.; Colby, K.A.; Skondra, D. Association of Metformin Use With Age-Related Macular Degeneration: A Case-Control Study. *JAMA Ophthalmol.* **2021**, *139*, 302–309. [CrossRef]
26. He, M.-S.; Chang, F.-L.; Lin, H.-Z.; Wu, J.-L.; Hsieh, T.-C.; Lee, Y.-C. The Association Between Diabetes and Age-Related Macular Degeneration Among the Elderly in Taiwan. *Diabetes Care* **2018**, *41*, 2202–2211. [CrossRef]
27. Abokyi, S.; To, C.H.; Lam, T.T.; Tse, D.Y. Central Role of Oxidative Stress in Age-Related Macular Degeneration: Evidence from a Review of the Molecular Mechanisms and Animal Models. *Oxid Med. Cell Longev.* **2020**, *2020*, 7901270. [CrossRef]
28. Tisi, A.; Feligioni, M.; Passacantando, M.; Ciancaglini, M.; Maccarone, R. The Impact of Oxidative Stress on Blood-Retinal Barrier Physiology in Age-Related Macular Degeneration. *Cells* **2021**, *10*, 64. [CrossRef]
29. Bloch-Damti, A.; Bashan, N. Proposed mechanisms for the induction of insulin resistance by oxidative stress. *Antioxid Redox Signal* **2005**, *7*, 1553–1567. [CrossRef]
30. Newsholme, P.; Keane, K.N.; Carlessi, R.; Cruzat, V. Oxidative stress pathways in pancreatic β-cells and insulin-sensitive cells and tissues: Importance to cell metabolism, function, and dysfunction. *Am. J. Physiol. Cell Physiol.* **2019**, *317*, C420–C433. [CrossRef]
31. Liu, X. Overstimulation can create health problems due to increases in PI3K/Akt/GSK3 insensitivity and GSK3 activity. *Springerplus* **2014**, *3*, 356. [CrossRef]
32. Strunz, T.; Lauwen, S.; Kiel, C.; Hollander, A.D.; Weber, B.H.F. A transcriptome-wide association study based on 27 tissues identifies 106 genes potentially relevant for disease pathology in age-related macular degeneration. *Sci. Rep.* **2020**, *10*, 1584. [CrossRef]
33. Lim, L.S.; Mitchell, P.; Seddon, J.M.; Holz, F.G.; Wong, T.Y. Age-related macular degeneration. *Lancet* **2012**, *379*, 1728–1738. [CrossRef]
34. Skyler, J.S. Diabetes mellitus: Pathogenesis and treatment strategies. *J. Med. Chem.* **2004**, *47*, 4113–4117. [CrossRef]

35. Mitchell, P.; Wang, J.J. Diabetes, fasting blood glucose and age-related maculopathy: The Blue Mountains Eye Study. *Aust. N. Z. J. Ophthalmol.* **1999**, *27*, 197–199. [CrossRef]
36. Srinivasan, S.; Swaminathan, G.; Kulothungan, V.; Ganesan, S.; Sharma, T.; Raman, R. Age-related macular degeneration in a South Indian population, with and without diabetes. *Eye* **2017**, *31*, 1176–1183. [CrossRef]
37. Song, S.O.; Lee, Y.-h.; Kim, D.W.; Song, Y.D.; Nam, J.Y.; Park, K.H.; Kim, D.J.; Park, S.W.; Lee, H.C.; Lee, B.-W. Trends in Diabetes Incidence in the Last Decade Based on Korean National Health Insurance Claims Data. *ENM* **2016**, *31*, 292–299. [CrossRef]
38. Park, K.H.; Song, S.J.; Lee, W.K.; Yoon, H.S.; Koh, H.J.; Kim, C.G.; Kim, Y.Y.; Kang, S.W.; Kim, H.K.; Lee, B.R.; et al. The Results of Nation-Wide Registry of Age-related Macular Degeneration in Korea. *J. Korean Ophthalmol. Soc.* **2010**, *51*, 516–523. [CrossRef]
39. Carneiro, Â.; Andrade, J.P. Nutritional and Lifestyle Interventions for Age-Related Macular Degeneration: A Review. *Oxid. Med. Cell Longev.* **2017**, *2017*, 6469138.
40. Nguyen, A.M.; Arora, K.S.; Swenor, B.K.; Friedman, D.S.; Ramulu, P.Y. Physical activity restriction in age-related eye disease: A cross-sectional study exploring fear of falling as a potential mediator. *BMC Geriatr.* **2015**, *15*, 64. [CrossRef]
41. Willis, J.R.; Jefferys, J.L.; Vitale, S.; Ramulu, P.Y. Visual impairment, uncorrected refractive error, and accelerometer-defined physical activity in the United States. *Arch. Ophthalmol.* **2012**, *130*, 329–335. [CrossRef]
42. van Landingham, S.W.; Willis, J.R.; Vitale, S.; Ramulu, P.Y. Visual field loss and accelerometer-measured physical activity in the United States. *Ophthalmology* **2012**, *119*, 2486–2492. [CrossRef]
43. Crews, J.E.; Chou, C.F.; Zhang, X.; Zack, M.M.; Saaddine, J.B. Health-related quality of life among people aged ≥65 years with self-reported visual impairment: Findings from the 2006–2010 behavioral risk factor surveillance system. *Ophthalmic Epidemiol.* **2014**, *21*, 287–296. [CrossRef]
44. Langelaan, M.; de Boer, M.R.; van Nispen, R.M.; Wouters, B.; Moll, A.C.; van Rens, G.H. Impact of visual impairment on quality of life: A comparison with quality of life in the general population and with other chronic conditions. *Ophthalmic Epidemiol.* **2007**, *14*, 119–126. [CrossRef]
45. Vu, H.T.; Keeffe, J.E.; McCarty, C.A.; Taylor, H.R. Impact of unilateral and bilateral vision loss on quality of life. *Br. J. Ophthalmol.* **2005**, *89*, 360–363. [CrossRef]
46. Crews, J.E.; Campbell, V.A. Vision impairment and hearing loss among community-dwelling older Americans: Implications for health and functioning. *Am. J. Public Health* **2004**, *94*, 823–829. [CrossRef]
47. Ramrattan, R.S.; Wolfs, R.C.; Panda-Jonas, S.; Jonas, J.B.; Bakker, D.; Pols, H.A.; Hofman, A.; de Jong, P.T. Prevalence and causes of visual field loss in the elderly and associations with impairment in daily functioning: The Rotterdam Study. *Arch. Ophthalmol.* **2001**, *119*, 1788–1794. [CrossRef]
48. Wallhagen, M.I.; Strawbridge, W.J.; Shema, S.J.; Kurata, J.; Kaplan, G.A. Comparative impact of hearing and vision impairment on subsequent functioning. *J. Am. Geriatr. Soc.* **2001**, *49*, 1086–1092. [CrossRef]
49. Bookwala, J.; Lawson, B. Poor vision, functioning, and depressive symptoms: A test of the activity restriction model. *Gerontologist* **2011**, *51*, 798–808. [CrossRef]
50. Klein, B.E.; Klein, R.; Knudtson, M.D.; Lee, K.E. Relationship of measures of frailty to visual function: The Beaver Dam Eye Study. *Trans. Am. Ophthalmol. Soc.* **2003**, *101*, 191–196; discussion 196–199.
51. Cupples, M.E.; Hart, P.M.; Johnston, A.; Jackson, A.J. Improving healthcare access for people with visual impairment and blindness. *BMJ* **2012**, *344*, e542. [CrossRef]
52. Jones, N.; Bartlett, H. The impact of visual impairment on nutritional status: A systematic review. *Br. J. Vis. Impair.* **2018**, *36*, 17–30. [CrossRef]
53. DeAngelis, M.M.; Owen, L.A.; Morrison, M.A.; Morgan, D.J.; Li, M.; Shakoor, A.; Vitale, A.; Iyengar, S.; Stambolian, D.; Kim, I.K.; et al. Genetics of age-related macular degeneration (AMD). *Hum. Mol. Genet.* **2017**, *26*, R45–R50. [CrossRef]
54. Spencer, K.L.; Glenn, K.; Brown-Gentry, K.; Haines, J.L.; Crawford, D.C. Population differences in genetic risk for age-related macular degeneration and implications for genetic testing. *Arch. Ophthalmol.* **2012**, *130*, 116–117. [CrossRef]
55. Swenor, B.K.; Simonsick, E.M.; Ferrucci, L.; Newman, A.B.; Rubin, S.; Wilson, V. Visual impairment and incident mobility limitations: The health, aging and body composition study. *J. Am. Geriatr. Soc.* **2015**, *63*, 46–54. [CrossRef]
56. Gorusupudi, A.; Nelson, K.; Bernstein, P.S. The Age-Related Eye Disease 2 Study: Micronutrients in the Treatment of Macular Degeneration. *Adv. Nutr.* **2017**, *8*, 40–53. [CrossRef]

biomedicines

MDPI

Article

Correlation between Macular Neovascularization (MNV) Type and Druse Type in Neovascular Age-Related Macular Degeneration (AMD) Based on the CONAN Classification

Daniel Rudolf Muth [1,2,*], Mario Damiano Toro [1,3,4], Anahita Bajka [1], Kamil Jonak [5,6], Roman Rieder [1], Myrtha Magdalena Kohler [1], Jeanne Martine Gunzinger [1], Eric H. Souied [7], Michael Engelbert [8,9], K. Bailey Freund [8,9] and Sandrine Anne Zweifel [1,*]

1 Department of Ophthalmology, University Hospital Zurich (USZ), University of Zurich (UZH), 8091 Zurich, Switzerland
2 University of Zurich (UZH), 8006 Zurich, Switzerland
3 Department of General and Pediatric Ophthalmology, Medical University of Lublin, 20-093 Lublin, Poland
4 Eye Clinic, Public Health Department, University of Naples Federico II, 80138 Naples, Italy
5 Department of Biomedical Engineering, Lublin University of Technology, 20-618 Lublin, Poland
6 Department of Psychiatry, Psychotherapy and Early Intervention, Medical University of Lublin, 20-093 Lublin, Poland
7 Department of Ophthalmology, Centre Hospitalier Intercommunal de Creteil, University Paris Est Creteil, 94000 Creteil, France
8 Vitreous Retina Macula Consultants of New York, New York, NY 10022, USA
9 Department of Ophthalmology, NYU Grossman School of Medicine, New York, NY 10016, USA
* Correspondence: dr-muth@t-online.de (D.R.M.); sandrine.zweifel@usz.ch (S.A.Z.); Tel.: +41-44-255-87-94 (D.R.M. & S.A.Z.); Fax: +41-44-255-44-72 (D.R.M. & S.A.Z.)

Citation: Muth, D.R.; Toro, M.D.; Bajka, A.; Jonak, K.; Rieder, R.; Kohler, M.M.; Gunzinger, J.M.; Souied, E.H.; Engelbert, M.; Freund, K.B.; et al. Correlation between Macular Neovascularization (MNV) Type and Druse Type in Neovascular Age-Related Macular Degeneration (AMD) Based on the CONAN Classification. *Biomedicines* **2022**, *10*, 2370. https://doi.org/10.3390/biomedicines10102370

Academic Editor: Oyuna Kozhevnikova

Received: 20 August 2022
Accepted: 13 September 2022
Published: 22 September 2022

Abstract: Abstract: BackgroundTo investigate associations and predictive factors between macular neovascularization (MNV) lesion variants and drusen types in patients with treatment-naïve neovascular age-related macular degeneration (AMD). Methods: Multimodal imaging was retrospectively reviewed for druse type (soft drusen, subretinal drusenoid deposits (SDDs) or mixed) and MNV type (MNV 1, MNV 2, MNV 1/2 or MNV 3). The Consensus on Neovascular AMD Nomenclature (CONAN) classification was used for characterizing MNV at baseline. Results: One eye of each eligible patient was included (n = 191). Patients with predominant SDDs had an increased adjusted odds ratio (aOR) for MNV 2 (23.4453, p = 0.0025) and any type of MNV 3 (8.7374, $p < 0.0001$). Patients with MNV 1/2 had an aOR for predominant SDDs (0.3284, p = 0.0084). Patients with MNV1 showed an aOR for SDDs (0.0357, $p < 0.0001$). Eyes with SDDs only without other drusen types showed an aOR for MNV 2 (9.2945, $p < 0.0001$). Conclusions: SDDs represent a common phenotypic characteristic in AMD eyes with treatment-naïve MNV. The aOR for eyes with predominant SDDs to develop MNV 2 and MNV 3 was much higher, possibly due to their location in the subretinal space. The predominant druse type may help to predict which type of MNV will develop during the course of AMD.

Keywords: subretinal drusenoid deposits; SDDs; reticular pseudodrusen; spectral domain optical coherence tomography; OCT; MNV; macular neovascularization; CNV; age-related macular degeneration; AMD

1. Introduction

With the availability of high-resolution retinal imaging, the classification of macular neovascularization (MNV) and drusen has evolved. Based on anatomical localization and multimodal imaging, including fluorescein angiography and spectral domain optical coherence tomography (SD-OCT), a revised classification scheme was proposed by Freund et al. [1]. This classification system built upon Grossniklaus' and Gass' original observations from histopathologic slides of neovascular age-related macular degeneration (AMD) differentiating between vessels confined to the sub-retinal pigment epithelium space, described as type 1 MNV, and vessels proliferating above the retinal pigment epithelium in the subneurosensory, subretinal space, described as type 2 MNV [2]. Gass had recognized that a distinction between type 1 and type 2 MNV based on biomicroscopic and fluorescein angiographic findings "is not always easy and in some cases impossible" [3]. Incorporating findings from fluorescein angiography and OCT for grading neovascular AMD, higher incidences of intraretinal neovascularization, described as type 3 MNV, and mixed types 1 and 2 were found compared with those reported in prior studies using fluorescein alone [4]. A new nomenclature was proposed by the Consensus on Neovascular Age-Related Macular Degeneration Nomenclature (CONAN) Study Group in 2020 [5].

Optical coherence tomography has improved our understanding of not only MNV types, but has also expanded and refined distinctions between drusen types. The clinical appearance of reticular pseudodrusen, first described in 1990 [6], could be linked to aggregations containing typical drusen-associated material located in the subretinal space, which were termed subretinal drusenoid deposits [7]. Subretinal drusenoid deposits have a high prevalence in AMD, which has been underestimated prior to the SD-OCT era [7]. Subretinal drusenoid deposits have been recognized as an additional feature of early and both forms of late AMD, called geographic atrophy/complete retinal pigment epithelium and outer retinal (cRORA) and neovascular AMD [8–10]. Late AMD is associated with vision loss [11]. The latest report of the Age-Related Eye Diseases Study 2 concluded that subretinal drusenoid deposits significantly contribute to the development of late AMD stages, especially in patients who already present subretinal drusenoid deposits at an early, low-disease-severity stage [11]. Subretinal drusenoid deposits mainly seemed to contribute to the development of geographic atrophy [11]. However, soft drusen and subretinal drusenoid deposits were also found to be risk factors for MNV [12]. There are published data suggesting that patients with subretinal drusenoid deposits are more likely to develop macular type 3 MNV [13,14].

The purpose of this study was to assess the predictive odds ratios between drusen type and MNV lesion variants in patients with treatment-naïve neovascular AMD [5].

2. Materials and Methods

2.1. Ethics

Ethics Committee approval was obtained from the Local Ethics Committee of the Canton of Zurich (approval number: PB_2016-00264). This study adheres to the tenets of the 1964 Declaration of Helsinki and its later amendments.

2.2. Study Design

This is a single-center, retrospective, observational study conducted at the Department of Ophthalmology of the University Hospital of Zurich (USZ), Switzerland.

2.3. Data Collection

All fluoresceine angiographies (FA) performed between 2011 and 2013 at the University Hospital of Zurich were screened for the presence of treatment-naïve macular neovascularization (MNV). Inclusion criteria for this study were the presence of treatment-naïve MNV secondary to AMD as evidenced by multimodal imaging, including FA, spectral-domain optical coherence tomography (SD-OCT), fundus autofluorescence (FAF) imaging and by clinical examination in patients aged ≥50 years. All neovascular lesion types, including polypoidal choroidal vasculopathy (PCV), were evaluated. Eyes with PCV were included only when structural signs of AMD were also present. Eyes with central serous chorioretinopathy (CSC) with or without PCV were excluded. In patients with no available baseline images of the study eye prior to the development of MNV, druse type was evaluated in the fellow eye provided there was no evidence of MNV. Only one eye per participant was selected and included in the study. If both eyes of the same patient were eligible, one eye was chosen randomly.

Exclusion criteria included patients with bilateral MNV at baseline, eyes with predominantly fibrotic lesions and those with such poor image quality that reliable classification of MNV and drusen was not possible.

Patients with bilateral MNV at baseline were excluded from the study as there would not be any pre-conversion images or a non-neovascular fellow eye to allow the grading of drusen type.

Neovascularizations that were confined to the space under the retinal pigment epithelium (RPE) were defined as MNV 1. PCV was considered an aneurysmal variant of MNV 1 and therefore subsumed under MNV 1. Neovascularizations proliferating above the RPE, in the subneurosensory, subretinal space, were defined as MNV 2. Cases with mixed type 1 and 2 lesions were recorded as a separate subgroup termed MNV 1/2. Cases with multiple lesion types including any type 3 lesion (MNV 1/3 and MNV 2/3) were summarized under MNV 3.

Descriptive statistics regarding the participant demographics are listed in Table 1.

Table 1. Demographic data.

Total patients screened (n_{screen})	1990 (3980 eyes)
Patients identified as suitable ($n_{patient}$)	268 (536 eyes)
Patients removed due to fibrovascular scarring or poor image quality ($n_{dropped}$)	77 (154 eyes) (29%)
Patients included ($n_{patient}$)	191 (382 eyes) (71%)
Eyes included (1 per patient) (n)	191 eyes
Localization OD/OS	97 (51%)/94 (49%)
Gender female/male	117 (61%)/74 (39%)
Age mean (years) ± SD	77.9 ± 7.8
BCVA mean ± SD	56.6 ± 21.2 ETDRS letters
Glaucoma	22 (12%)
Arterial hypertension	92 (48%)
Diabetes mellitus	17 (9%)

Legend: OD, oculus dexter; OS, oculus sinister; SD, standard deviation of the arithmetic mean; BCVA, best-corrected visual acuity, ETDRS, Early Treatment Diabetic Retinopathy Study.

FA images were obtained with either a fundus camera system (Carl Zeiss AG, Oberkochen, Germany) or the Heidelberg Viewing Module (version 6.0.9.0) included in our Spectralis SD-OCT device (version 1.9.10.0; Heidelberg Engineering GmbH, Heidelberg, Germany). Indocyanine green angiography (ICGA) and FAF images were obtained from all patients using the confocal scanning laser ophthalmoscope (CSLO) (Heidelberg Retina Angiograph, HRA2, Heidelberg Engineering, Heidelberg, Germany).

2.4. Image Grading and Analysis

Neovascular lesions were subtyped according to CONAN criteria [1,5]. See Figures 1 and 2 for examples [1,4].

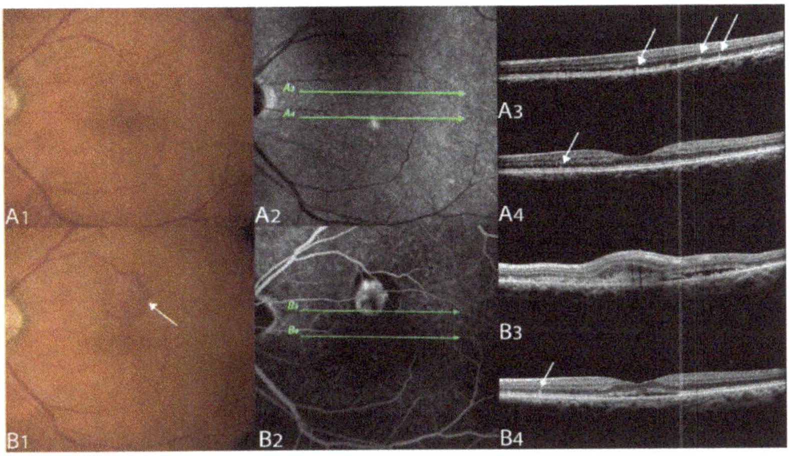

Figure 1. Multimodal imaging example of a type 2 macular neovascularization.

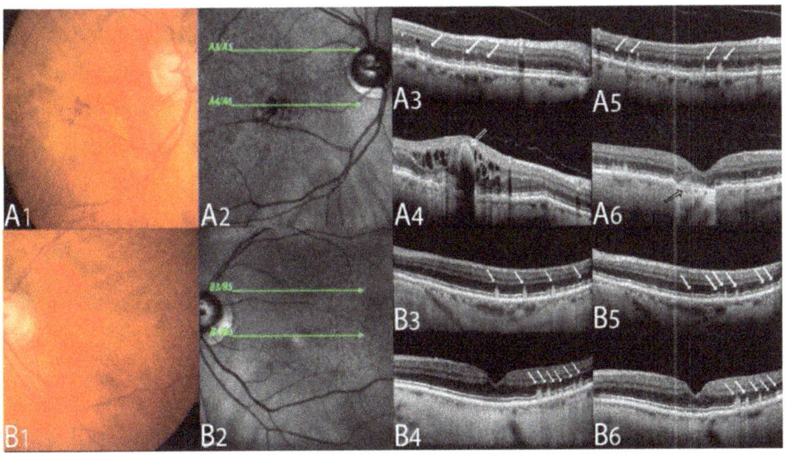

Figure 2. Multimodal imaging example of a type 3 macular neovascularization.

Drusen types were categorized into soft drusen and SDDs. SDDs were considered present when there was OCT evidence of ≥5 definite SDDs above the RPE in >1 B-scan, with consistent changes in either the near-infrared imaging or the blue light channel, as previously described [7]. Soft drusen were determined from color fundus photographs and confirmed by SD-OCT. If both types of drusen were present, the predominance of either soft drusen or SDDs was determined by two experienced graders; in the case of discordance among the readers, a senior retina specialist was consulted (SAZ) (see Table 2).

Table 2. Descriptive statistics of drusen types and MNV types.

	Total	MNV 1	MNV 2	MNV 3	MNV Mixed 1/2
Eyes included (*n*)	191 (100%)	80 (42%)	23 (12%)	49 (26%)	39 (20%)
Soft drusen only	38 (20%)	34 (43%)	0 (0%)	1 (2%)	3 (8%)
Predominant soft drusen	43 (23%)	32 (40%)	1 (4%)	4 (8%)	6 (15%)
SDD only	39 (20%)	3 (4%)	14 (61%)	13 (27%)	9 (23%)
Predominant SDD	71 (37%)	11 (14%)	8 (35%)	31 (63%)	21 (54%)

Legend: *n*, number of included eyes; SDD, subretinal drusenoid deposit; MNV, macular neovascularization.

The druse type that occupied the greater number of subfields in the Early Treatment Diabetic Retinopathy Study (ETDRS) macular grid was determined as the predominant druse type.

2.5. Statistical Analysis

Data were organized in Microsoft Excel (Microsoft Corp., Redmond, WA, USA) and statistically analyzed using SPSS software version 23 (SPSS Inc., Chicago, IL, USA) and R.app v4.1.0 GUI 1.76 for MacOS (The R Foundation for Statistical Computing c/o Institute for Statistics and Mathematics, 1020 Vienna, Austria). Descriptive statistics, such as the mean and standard deviation (SD), were computed. The number of neovascular AMD lesions as identified by the anatomic classification system was recorded.

Adjusted odds ratios (aORs), as a predictor between druse type and MNV type, were calculated in R using binary logistic regression. As the characteristic structural changes in AMD are age-related, age will not be normally distributed in our dataset. The statistical feature "age" will additionally contribute to the odds ratios (ORs). As we wanted to evaluate the effect of drusen on MNV development alone, we adjusted the ORs for age. The statistical significance level (α) was defined as 0.05. Statistical analysis results with a p-value less than 0.05 were interpreted as statistically significant.

3. Results

A total of 3980 eyes with treatment-naïve neovascular AMD were screened. Among these, 536 eyes met the eligibility criteria. Of those, 154 eyes were excluded due to advanced fibrovascular scarring and/or poor image quality precluding a reliable classification of MNV and drusen type. Eventually, 382 eyes of 191 patients were eligible. Only 1 eye per patient was included, resulting in 191 study eyes (n = 191) (Table 1). If both eyes were eligible, one eye was chosen randomly. Patients with the predominant druse type of SDD had higher aORs for MNV 2 and MNV 3 (Table 3). Patients with MNV 1 were less likely to have predominant SDDs (Table 3). No clear predictor for any druse type was found for mixed MNV 1/2 lesions. Patients with soft drusen showed the highest overall aOR with MNV 1 lesion (Table 3).

History of diabetes, arterial hypertension and glaucoma did not show statistically significant correlations (Phi/Cramer-v) with neither MNV type nor druse type (Table 4).

Table 3. Predictors (aOR adjusted for age) between druse and MNV type.

	aOR	95% CI	*p*
		MNV 1	
Soft drusen only	19.0008	7.0019; 66.9073	**<0.0001**
Predominant soft drusen	27.9873	13.0180; 64.5909	**<0.0001**
SDD only	0.0806	0.01867; 0.2395	**<0.0001**
Predominant SDD	0.0357	0.0155; 0.0768	**<0.0001**
		MNV 2	
Soft drusen only	n.s.	n.s.	0.8700
Predominant soft drusen	0.0427	0.0023; 0.2165	**0.0025**
SDD only	9.2945	3.6536; 24.9841	**<0.0001**
Predominant SDD	23.4453	4.6190; 429.4434	**0.0025**
		MNV 3	
Soft drusen only	0.0661	0.0037; 0.3243	**0.00853**
Predominant soft drusen	0.1145	0.0376; 0.2849	**<0.0001**
SDD only	n.s.	n.s.	0.3198
Predominant SDD	8.7374	0.5105; 26.5916	**<0.0001**
		MNV 1/2 mixed	
Soft drusen only	0.2816	0.0650; 0.8484	**0.0455**
Predominant soft drusen	0.3284	0.1364; 0.7268	**0.0084**
SDD only	n.s.	n.s.	0.6690
Predominant SDD	0.3284	0.1364; 0.7268	**0.0084**

Legend: 95% CI, 95% confidence interval; aOR, adjusted odds ratio; MNV, macular neovascularization; NA, not applicable; n.s., not statistically significant; *p*, probability of binary logistic regression (defined significant when $p < 0.05$; SDD, subretinal drusenoid deposit.

Table 4. Correlations (Phi/Cramer-V) between MNV type and druse type with systemic factors.

	Coeff.	95% CI	*p*
MNV type vs. Glaucoma	0.0757	0.0000; 0.0966	0.9063
MNV type vs. Hypertension	0.1153	0.0000; 0.1677	0.5471
MNV type vs. Diabetes	0.0726	0.0000; 0.0887	0.9221
Druse type vs. Glaucoma	0.0788	0.0000; 0.1048	0.8828
Druse type vs. Hypertension	0.1091	0.0000; 0.1321	0.6556
Druse type vs. Diabetes	0.1269	0.0000; 0.1591	0.4171

Legend: 95% CI, 95% confidence interval; Coeff., correlation coefficient of Phi/Cramer-V correlation for categorial/nominal data; MNV, macular neovascularization; n.s., not statistically significant; *p*, Pearson's Chi-squared *p*-value (defined as significant when $p < 0.05$); SDD, subretinal drusenoid deposit.

4. Discussion

The subretinal space is an unusual location for the extracellular deposition of material, and not only changes in the sub-RPE location, but also changes anterior to the RPE cells, may play an important role in AMD pathogenesis. The presence of SDDs seems to be associated with thin choroidal thickness [15,16]. Eyes with sub-RPE predominant soft drusen show the highest overall odds ratio for developing neovascular AMD [17]. We observed the same trend in our data (Table 3). However, subretinal SDDs clearly seem to be associated with late AMD, including complete outer retinal atrophy (cRORA) as well as MNV in AMD [11,18–20]. There seemed to be debate as to whether SDDs could be linked to an AMD MNV subtype [21]. Cohen et al. were the first to observe a correlation between SDD and the MNV 3 subtype [13]. Marsiglia et al. demonstrated that patients with MNV 1 were less likely to have SDDs and that patients with MNV 3 were more likely to have SDDs in their non-neovascular fellow eye [14]. They did not observe an association between SDDs and MNV 2, which might be due to the low number of only 10 eyes with MNV 2 in their series [14]. Applying the CONAN Study Group criteria for MNV classification to this dataset, we could demonstrate that, in AMD patients with treatment-naïve MNV, SDDs were more likely to be a risk factor for MNV 2 and MNV 3 as compared with soft drusen (Table 3). MNV 2 occurs in the subretinal space above the RPE. MNV 3 originates intraretinally (supra-RPE) and penetrates the RPE only in later stages. Therefore, SDDs seem to be a plausible risk factor for the development of both the MNV 2 and MNV 3 subtypes. This association is supported by findings from Spaide et al., Rabiolo et al. and Lee et al., who demonstrated that SDDs would preferably progress to MNV 2 and MNV 3 [16,19,20,22]. We could confirm this finding with MNV 2 and with MNV 3 for predominant SDDs. Interestingly, we did not find a statistically significant odds ratio for SDDs only with any type of MNV 3. This might be due to our classification of the presence of any MNV 3 under one group. Hence, the MNV 3 group in our series also included mixed lesions, such as MNV 1/3 and MNV 2/3, making it difficult to find a statistical predictor.

Ahmed et al. frequently discovered MNV 2 without the presence of extracellular deposits [23]. This highlights that SDDs may be difficult to reliably detect. Rabiolo et al. also recommended using multimodal imaging with at least two different modalities to reliably detect SDDs [22].

The limitations of this study include its retrospective nature and the fact that graders were not masked to the original diagnosis of neovascular AMD. The non-neovascular fellow eye was used to identify the type of drusen in patients with no available baseline images of the study eye prior to the development of MNV; however, this is supported by prior studies showing intraindividual symmetry in eyes with AMD [24–27].

5. Conclusions

The anatomical classification of MNV, which incorporates FA and SD-OCT findings, seems to provide more accurate information about associations between the type of drusen and type of MNV. Predicting the type of MNV for patients with intermediate AMD might inform more personalized patient care with respect to monitoring and future treatment.

Author Contributions: Conceptualization: S.A.Z.; Methodology: S.A.Z., D.R.M. and K.J.; Software: D.R.M. and K.J.; Validation: S.A.Z., D.R.M., K.J., E.H.S., M.E., K.B.F. and M.D.T.; Formal analysis: S.A.Z., D.R.M., K.J., E.H.S., M.E., K.B.F. and M.D.T.; Investigation: S.A.Z., D.R.M., M.D.T., A.B., R.R., M.M.K. and J.M.G.; Resources: S.A.Z. and D.R.M.; Data curation: S.A.Z., D.R.M., K.J., A.B. and R.R., M.M.K. and J.M.G.; Writing—Original draft preparation: S.A.Z. and D.R.M.; Writing—Review and editing: S.A.Z., D.R.M., E.H.S., M.E. and K.B.F.; Visualization: D.R.M. and K.J.; Supervision: S.A.Z. and D.R.M.; Project administration: S.A.Z.; Funding acquisition: N/A. All authors have read and agreed to the published version of the manuscript.

Funding: This research received no external funding.

Institutional Review Board Statement: Ethics Committee approval was obtained from the Local Ethics Committee of the Canton of Zurich (approval number: PB_2016-00264). This study adheres to the tenets of the 1964 Declaration of Helsinki and its later amendments.

Informed Consent Statement: Informed consent was obtained from all subjects involved in the study.

Data Availability Statement: The data, aside from the data published in this manuscript, are not publicly available due to privacy restrictions.

Conflicts of Interest: K.B.F.: Consultant—Allergan, Bayer HealthCare, Carl Zeiss Meditec, Genentech, Inc., Heidelberg Engineering, Novartis, Regeneron; Research support—Genentech, Inc., Roche. S.A.Z.: Consultant—Allergan, Bayer HealthCare, Novartis, Roche; Grant Support—Bayer HealthCare, Novartis. None of the disclosures are relevant for this manuscript.

References

1. Freund, K.B.; Zweifel, S.A.; Engelbert, M. Do we need a new classification for choroidal neovascularization in age-related macular degeneration? *Retina* **2010**, *30*, 1333–1349. [CrossRef] [PubMed]
2. Grossniklaus, H.E.; Gass, J.D. Clinicopathologic correlations of surgically excised type 1 and type 2 submacular choroidal neovascular membranes. *Am. J. Ophthalmol.* **1998**, *126*, 59–69. [CrossRef]

3. Gass, J.D. Biomicroscopic and histopathologic considerations regarding the feasibility of surgical excision of subfoveal neovascular membranes. *Am. J. Ophthalmol.* **1994**, *118*, 285–298. [CrossRef]
4. Jung, J.J.; Chen, C.Y.; Mrejen, S.; Gallego-Pinazo, R.; Xu, L.; Marsiglia, M.; Boddu, S.; Freund, K.B. The incidence of neovascular subtypes in newly diagnosed neovascular age-related macular degeneration. *Am. J. Ophthalmol.* **2014**, *158*, 769–779 e762. [CrossRef] [PubMed]
5. Spaide, R.F.; Jaffe, G.J.; Sarraf, D.; Freund, K.B.; Sadda, S.R.; Staurenghi, G.; Waheed, N.K.; Chakravarthy, U.; Rosenfeld, P.J.; Holz, F.G.; et al. Consensus Nomenclature for Reporting Neovascular Age-Related Macular Degeneration Data: Consensus on Neovascular Age-Related Macular Degeneration Nomenclature Study Group. *Ophthalmology* **2020**, *127*, 616–636. [CrossRef]
6. Mimoun, G.; Soubrane, G.; Coscas, G. Macular drusen. *J. Fr. Ophtalmol.* **1990**, *13*, 511–530.
7. Zweifel, S.A.; Spaide, R.F.; Curcio, C.A.; Malek, G.; Imamura, Y. Reticular pseudodrusen are subretinal drusenoid deposits. *Ophthalmology* **2010**, *117*, 303–312 e301. [CrossRef]
8. Schmitz-Valckenberg, S.; Alten, F.; Steinberg, J.S.; Jaffe, G.J.; Fleckenstein, M.; Mukesh, B.N.; Hohman, T.C.; Holz, F.G.; Geographic Atrophy Progression (GAP) Study Group. Reticular drusen associated with geographic atrophy in age-related macular degeneration. *Investig. Ophthalmol. Vis. Sci.* **2011**, *52*, 5009–5015. [CrossRef]
9. Steinberg, J.S.; Auge, J.; Jaffe, G.J.; Fleckenstein, M.; Holz, F.G.; Schmitz-Valckenberg, S.; Group, G.A.P.S. Longitudinal analysis of reticular drusen associated with geographic atrophy in age-related macular degeneration. *Investig. Ophthalmol. Vis. Sci.* **2013**, *54*, 4054–4060. [CrossRef]
10. Zhou, Q.; Daniel, E.; Maguire, M.G.; Grunwald, J.E.; Martin, E.R.; Martin, D.F.; Ying, G.S.; Comparison of Age-Related Macular Degeneration Treatments Trials Research Group. Pseudodrusen and Incidence of Late Age-Related Macular Degeneration in Fellow Eyes in the Comparison of Age-Related Macular Degeneration Treatments Trials. *Ophthalmology* **2016**, *123*, 1530–1540. [CrossRef]
11. Agron, E.; Domalpally, A.; Cukras, C.A.; Clemons, T.E.; Chen, Q.; Lu, Z.; Chew, E.Y.; Keenan, T.D.L.; Areds; Groups, A.R. Reticular Pseudodrusen: The Third Macular Risk Feature for Progression to Late Age-related Macular Degeneration: Age-Related Eye Disease Study 2 Report 30. *Ophthalmology*, 2022; *in press*. [CrossRef] [PubMed]
12. Zweifel, S.A.; Imamura, Y.; Spaide, T.C.; Fujiwara, T.; Spaide, R.F. Prevalence and significance of subretinal drusenoid deposits (reticular pseudodrusen) in age-related macular degeneration. *Ophthalmology* **2010**, *117*, 1775–1781. [CrossRef] [PubMed]
13. Cohen, S.Y.; Dubois, L.; Tadayoni, R.; Delahaye-Mazza, C.; Debibie, C.; Quentel, G. Prevalence of reticular pseudodrusen in age-related macular degeneration with newly diagnosed choroidal neovascularisation. *Br. J. Ophthalmol.* **2007**, *91*, 354–359. [CrossRef] [PubMed]
14. Marsiglia, M.; Boddu, S.; Chen, C.Y.; Jung, J.J.; Mrejen, S.; Gallego-Pinazo, R.; Freund, K.B. Correlation between neovascular lesion type and clinical characteristics of nonneovascular fellow eyes in patients with unilateral, neovascular age-related macular degeneration. *Retina* **2015**, *35*, 966–974. [CrossRef] [PubMed]
15. Cheung, C.M.G.; Gan, A.; Yanagi, Y.; Wong, T.Y.; Spaide, R. Association between Choroidal Thickness and Drusen Subtypes in Age-Related Macular Degeneration. *Ophthalmol. Retina* **2018**, *2*, 1196–1205. [CrossRef]
16. Lee, J.; Kim, M.; Lee, C.S.; Kim, S.S.; Koh, H.J.; Lee, S.C.; Byeon, S.H. Drusen Subtypes and Choroidal Characteristics in Asian Eyes with Typical Neovascular Age-Related Macular Degeneration. *Retina* **2020**, *40*, 490–498. [CrossRef]
17. Lee, J.; Choi, S.; Lee, C.S.; Kim, M.; Kim, S.S.; Koh, H.J.; Lee, S.C.; Byeon, S.H. Neovascularization in Fellow Eye of Unilateral Neovascular Age-related Macular Degeneration According to Different Drusen Types. *Am. J. Ophthalmol.* **2019**, *208*, 103–110. [CrossRef]
18. Wightman, A.J.; Guymer, R.H. Reticular pseudodrusen: Current understanding. *Clin. Exp. Optom.* **2019**, *102*, 455–462. [CrossRef]
19. Spaide, R.F.; Ooto, S.; Curcio, C.A. Subretinal drusenoid deposits AKA pseudodrusen. *Surv. Ophthalmol.* **2018**, *63*, 782–815. [CrossRef]
20. Spaide, R.F. IMPROVING THE AGE-RELATED MACULAR DEGENERATION CONSTRUCT: A New Classification System. *Retina* **2018**, *38*, 891–899. [CrossRef]
21. Wilde, C.; Patel, M.; Lakshmanan, A.; Morales, M.A.; Dhar-Munshi, S.; Amoaku, W.M. Prevalence of reticular pseudodrusen in eyes with newly presenting neovascular age-related macular degeneration. *Eur. J. Ophthalmol.* **2016**, *26*, 128–134. [CrossRef] [PubMed]
22. Rabiolo, A.; Sacconi, R.; Cicinelli, M.V.; Querques, L.; Bandello, F.; Querques, G. Spotlight on reticular pseudodrusen. *Clin. Ophthalmol.* **2017**, *11*, 1707–1718. [CrossRef]
23. Ahmed, D.; Stattin, M.; Haas, A.M.; Graf, A.; Krepler, K.; Ansari-Shahrezaei, S. Drusen characteristics of type 2 macular neovascularization in age-related macular degeneration. *BMC Ophthalmol.* **2020**, *20*, 381. [CrossRef]
24. Pauleikhoff, D.; Radermacher, M.; Spital, G.; Muller, C.; Brumm, G.; Lommatzsch, A.; Bird, A.C. Visual prognosis of second eyes in patients with unilateral late exudative age-related macular degeneration. *Graefes Arch. Clin. Exp. Ophthalmol.* **2002**, *240*, 539–542. [CrossRef]
25. Lavin, M.J.; Eldem, B.; Gregor, Z.J. Symmetry of disciform scars in bilateral age-related macular degeneration. *Br. J. Ophthalmol.* **1991**, *75*, 133–136. [CrossRef] [PubMed]

26. Chang, B.; Yannuzzi, L.A.; Ladas, I.D.; Guyer, D.R.; Slakter, J.S.; Sorenson, J.A. Choroidal neovascularization in second eyes of patients with unilateral exudative age-related macular degeneration. *Ophthalmology* **1995**, *102*, 1380–1386. [CrossRef]
27. Mann, S.S.; Rutishauser-Arnold, Y.; Peto, T.; Jenkins, S.A.; Leung, I.; Xing, W.; Bird, A.C.; Bunce, C.; Webster, A.R. The symmetry of phenotype between eyes of patients with early and late bilateral age-related macular degeneration (AMD). *Graefes Arch. Clin. Exp. Ophthalmol.* **2011**, *249*, 209–214. [CrossRef]

biomedicines

MDPI

Case Report

Bacillary Layer Detachment in Neovascular Age-Related Macular Degeneration: Case Series

Filomena Palmieri [1,2,*], Saad Younis [1,2], Walid Raslan [1,2] and Lorenzo Fabozzi [1,2,3]

1 Western Eye Hospital, Imperial College Healthcare NHS Trust, London NW1 5QH, UK
2 Surgery, Cancer and Cardiovascular Division, Ophthalmology Department, Imperial College Healthcare NHS Trust, London W2 1NY, UK
3 Moorfields Eye Hospital NHS Foundation Trust, London EC1V 2PD, UK
* Correspondence: filomena.palmieri@nhs.net

Abstract: Purpose: This study seeks to report the clinical and multimodal imaging findings of eight eyes of seven patients with neovascular age-related macular degeneration (nAMD) who developed bacillary layer detachment (BALAD). Setting/Venue: The patients were analysed at the Western Eye Hospital in London, UK. Methods: The approaches of this research include clinical examinations and multimodal imaging-based description of cases of nAMD with BALAD. Results: We report multimodal imaging findings of bacillary layer detachment (BALAD) in patients with nAMD. Conclusions: A bacillary layer detachment was detected in patients with neovascular age-related macular degeneration. This multimodal imaging finding is not commonly described in the literature for this disease.

Keywords: bacillary layer detachment; age-related macular degeneration; neovascular AMD; anti-VEGF; retinal haemorrhage; myoid; macula; retina; intravitreal injections

Citation: Palmieri, F.; Younis, S.; Raslan, W.; Fabozzi, L. Bacillary Layer Detachment in Neovascular Age-Related Macular Degeneration: Case Series. *Biomedicines* **2023**, *11*, 988. https://doi.org/10.3390/biomedicines11030988

Academic Editor: Oyuna Kozhevnikova

Received: 18 February 2023
Revised: 12 March 2023
Accepted: 14 March 2023
Published: 22 March 2023

1. Introduction

Bacillary layer detachment (BALAD) is a definition introduced for the first time in 2018 by Mehta et al. to describe an optical coherence tomography (OCT) finding in a patient with Toxoplasmosis chorioretinitis and pachychoroid disease [1]. It is characterized by a splitting of the photoreceptor layer at the myoid level, resulting in a space posterior to the external limiting membrane (ELM). Several authors have reported BALAD in various ophthalmic conditions as well as ocular trauma, multifocal placoid pigment epitheliopathy, pachychoroid syndrome, tubercular choroidal granuloma, acute idiopathic maculopathy, type 2 macular telangiectasia, Vogt–Koyanagi–Harada disease, choroidal metastasis, osteoma, and age-related macular degeneration (AMD) [1–20]. The exact BALAD physiopathology is unknown. However, inflammation, vascular changes, and choroidal thickening have been hypothesized as possible underlying mechanisms [2,5,7,11]. This study reports multimodal imaging findings of BALAD in patients with neovascular age-related macular degeneration (nAMD).

2. Methods

This study was conducted at Western Eye Hospital in London. Written informed consent was provided by all patients before participating in the study.

All the patients received a complete ophthalmic examination, including best corrected visual acuity (BCVA) with refraction using an Early Treatment Diabetic Retinopathy Study (ETDRS) chart, intraocular pressure (IOP) measurement measured via an iCARE tonometer, slit lamp evaluation, clinical examination of the posterior pole with a 66D or 90D indirect fundus-viewing lens, optical coherence tomography (OCT) and optical coherence tomography angiography (OCTA) of the macula region using the Spectralis HRA+OCT platform (Heidelberg Engineering, Heidelberg, Germany), and fundus autofluorescence (FAF) using

Optos Ultra-widefield (Nikon Co., Ltd., Tokyo, Japan) or Spectralis HRA+OCT platform (Heidelberg Engineering, Heidelberg, Germany), at baseline. The medical history, ocular history, and a list of concomitant medications were collected before starting treatment.

The data used were obtained through Medisoft EMR software (Medisoft Limited, Leeds, UK).

At each follow-up visit, all patients had their BCVA and IOP measured, slit lamp evaluation, fundus oculi examination with a 66D or 90D indirect fundus-viewing lens, and OCT scan of the macular region.

3. Findings

3.1. Case 1

An 85-year-old male patient presented to our facility in April 2022, complaining of a left eye vision drop. The patient was on Timolol eye drops for glaucoma.

The BCVA was 0.1 and 1.0 LogMAR in the right and left eye, respectively. The anterior segment examination was unremarkable in both eyes.

Fundus examination showed macular drusen in the right eye and a macular yellowish elevated lesion with retinal haemorrhage in the left eye (Figure 1A). Autofluorescence imaging (FAF) showed a hypoautofluorescent macular lesion in the left eye (Figure 1B). An OCT scan was performed, showing a dry macula with drusen in the right eye and a type 2 macular neovascularization (MNV) with subretinal fluid and splitting at the ellipsoid zone (EZ) creating a hypo-reflective cavity in the left eye; a hyper-reflective band, continuous with the EZ layer, was observed on the floor of the BALAD (Figure 1C).

The optical coherence tomography angiography (OCTA) scan confirmed the presence of MNV in the left eye (Figure 2). Based on clinical findings and multimodal imaging, the patient was diagnosed with left nAMD. Anti-vascular endothelial growth factor (VEGF) intravitreal injections were commenced, according to a treat and extend (T&E) regimen. To date, the patient received four intravitreal injections. Since the first injection, the OCT scan showed a complete resolution of the BALAD (Figure 1D). However, a retinal pigment epithelium (RPE) tear was noted (Figure 1E). The BCVA did not improve (1.5 logMAR on the last examination).

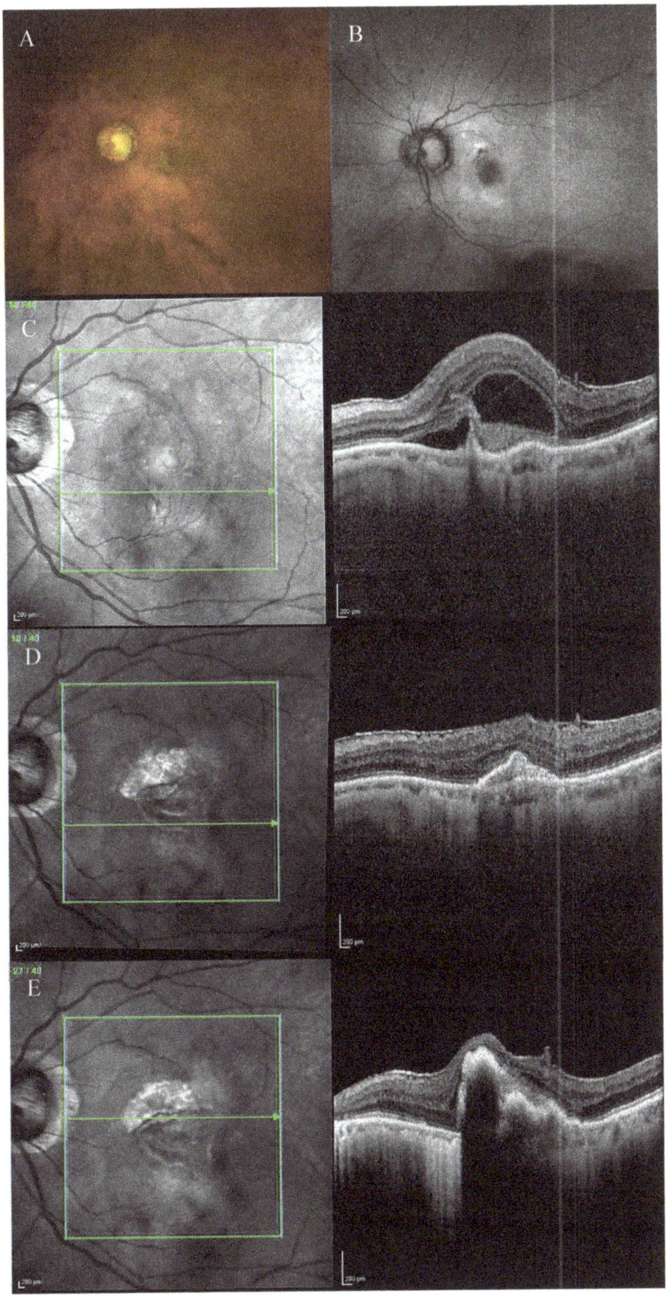

Figure 1. Case 1. Multimodal imaging features of bacillary layer detachment (BALAD) in a patient with left eye neovascular age-related macular degeneration. (**A**) Optos Ultra-widefield colour shows

a macular yellowish elevated lesion with subretinal haemorrhage. (**B**) Optos Ultra-widefield autofluorescence imaging shows a hypoautofluorescent macular lesion. (**C**). *Heidelberg* Spectralis optical coherence tomography shows a type 2 macular neovascularization with splitting at the ellipsoid zone creating an hypo-reflective cavity and a hyper-reflective band on the floor of the BALAD. (**D**) Optical coherence tomography shows a complete resolution of BALAD after four anti-vascular endothelial growth factor intravitreal injections. (**E**) *Heidelberg* Spectralis optical coherence tomography shows a retinal pigment epithelial tear.

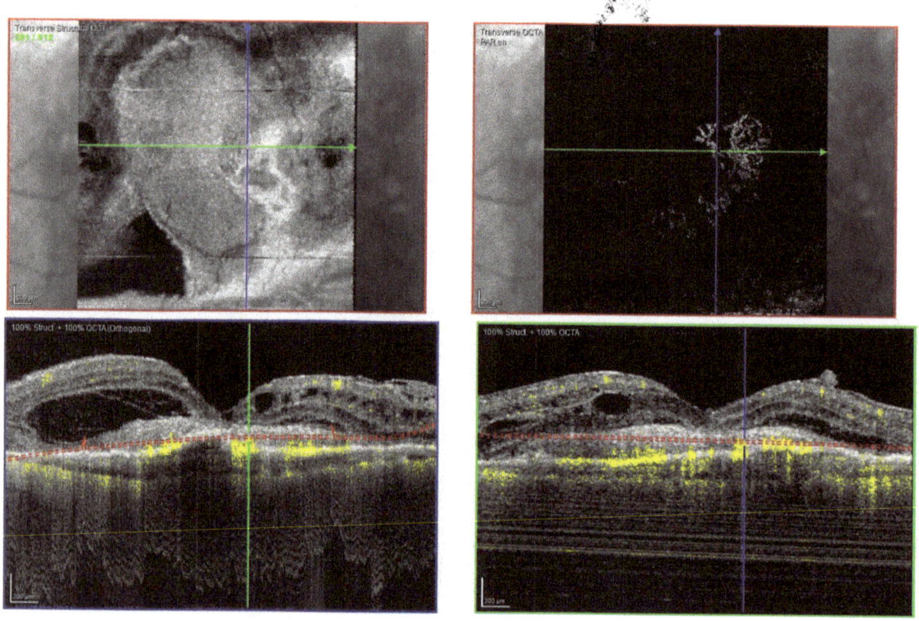

Figure 2. Case 1. Heidelberg Spectralis optical coherence tomography angiography shows a subretinal neovascular membrane in the left eye.

3.2. *Case 2*

A 79-year-old male patient with a pigmented lesion along the vascular inferotemporal arcade presented to our care in August 2017 for the first time.

The anterior segment was within normal limits in both eyes.

The OCT scan showed a macular choroidal lesion with subretinal fluid in the right eye and macular drusen in the left eye. A right eye ultrasound scan was also performed, showing a macular choroidal nevus. His BCVA was 0.7 and 0.00 LogMAR in the right and left eye, respectively.

The right eye lesion was monitored for more than a year until the patient experienced a sudden right eye vision drop. His right eye BCVA was counting fingers (CF).

Fundus examination showed a right macular yellowish elevated lesion with retinal haemorrhages on the superior border of the naevus (Figure 3A). FAF showed a hypoautofluorescent macular lesion (Figure 3B).

An OCT scan was performed showing a type 1 MNV with a haemorrhagic BALAD on the superior border of the nevus. The BALAD cavitation showed hyper-reflective material. Hyper-reflective granular foci at the ceiling of the BALAD were reported. The ELM was visible anterior to the BALAD (Figure 3C).

The OCTA scan confirmed the presence of the right MNV.

Neovascular AMD was diagnosed. The patient started anti-VEGF intravitreal injections according to a T&E protocol. After four injections, the BALAD showed a partial response to the treatment (Figure 3D). The BCVA improved to 0.8 logMAR.

However, the patient decided to stop the treatment for personal reasons.

In the following 8 months, follow-up visits were performed. A progressive increase in intraretinal fluid, subretinal hyper-reflective material, and subretinal fluid were observed (Figure 3E).

The BCVA remained stable.

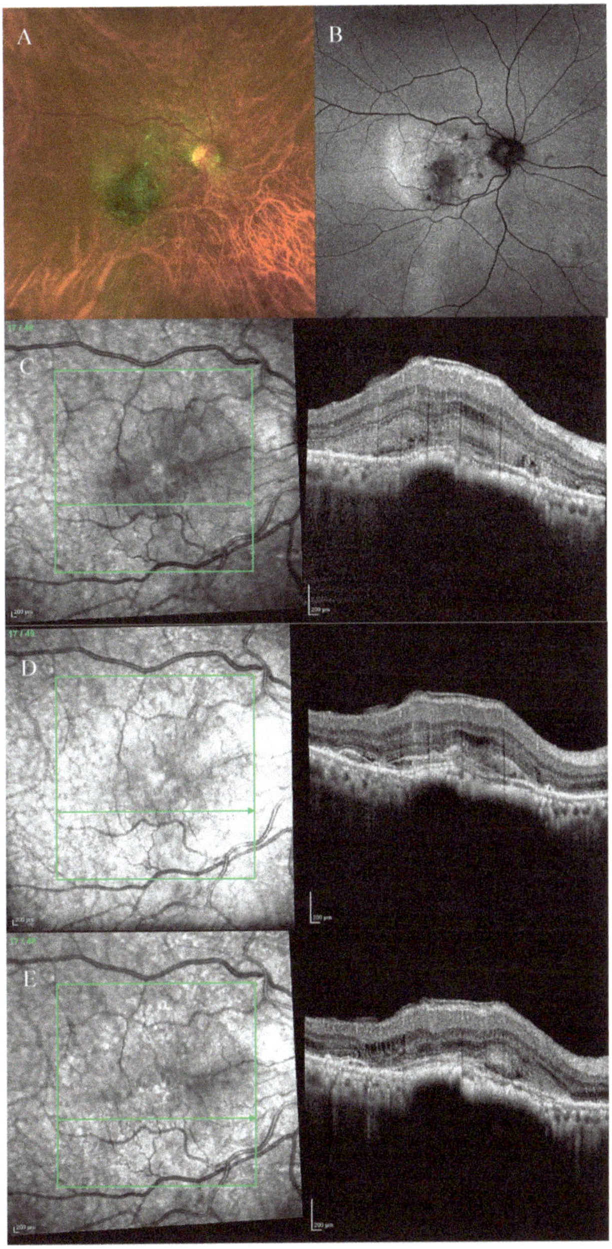

Figure 3. Case 2. Multimodal imaging features of bacillary layer detachment (BALAD) in a patient with right eye neovascular age-related macular degeneration on a choroidal naevus. (**A**) Optos Ultra-

widefield colour shows a macular yellowish elevated lesion with intraretinal haemorrhages on the superior border of a pigmented lesion. (**B**) Optos Ultra-widefield autofluorescence imaging shows a hypoautofluorescent macular lesion. (**C**) Heidelberg Spectralis optical coherence tomography shows a type 1 macular neovascularization with BALAD. The BALAD cavitation showed hyper-reflective material. The external limiting membrane is visible anterior to the BALAD. (**D**) Heidelberg Spectralis optical coherence tomography shows an improvement in BALAD after 4 intravitreal injections. (**E**) Heidelberg Spectralis optical coherence tomography shows increased intraretinal fluid, subretinal hyper-reflective material, and subretinal fluid eight months after discontinuing the treatment.

3.3. *Case 3*

A 62-year-old male patient presented to our care in December 2020 with a drop of vision in the left eye. His BCVA was 0.00 and 0.3 logMAR in the right and left eye, respectively.

The anterior segment examination was unremarkable in both eyes. An ophthalmological assessment showed features of right eye dry AMD and left eye nAMD with BALAD.

Fundus examination showed right macular drusen and a left macular subretinal fluid with a round yellow border (Figure 4A). FAF was also performed, showing hypoautofluorescence (Figure 4C).

The OCT scan showed a right dry macula with drusen and a left macular type 1 MNV with a fibrovascular pigment epithelial detachment (PED), subretinal fluid, and a split at the myoid zone creating an intraretinal space with multiple hyper-reflective foci and a septa-like arrangement. A hyper-reflective granular band at the anterior border of the BALAD was reported. The posterior border showed a hyper-reflective thickened band. The ELM was intact, recognizable, and anterior to the BALAD (Figure 4D,E).

The presence of the left MNV was confirmed by an OCTA scan.

An anti-VEGF intravitreal injection T&E regimen was commenced. However, the patient was non-compliant.

In March 2022, the patient decided to stop the treatment.

During the T&E treatment, the left eye developed subretinal fibrosis (Figure 4B) and no signs of visual acuity improvement were noticed in the following visits.

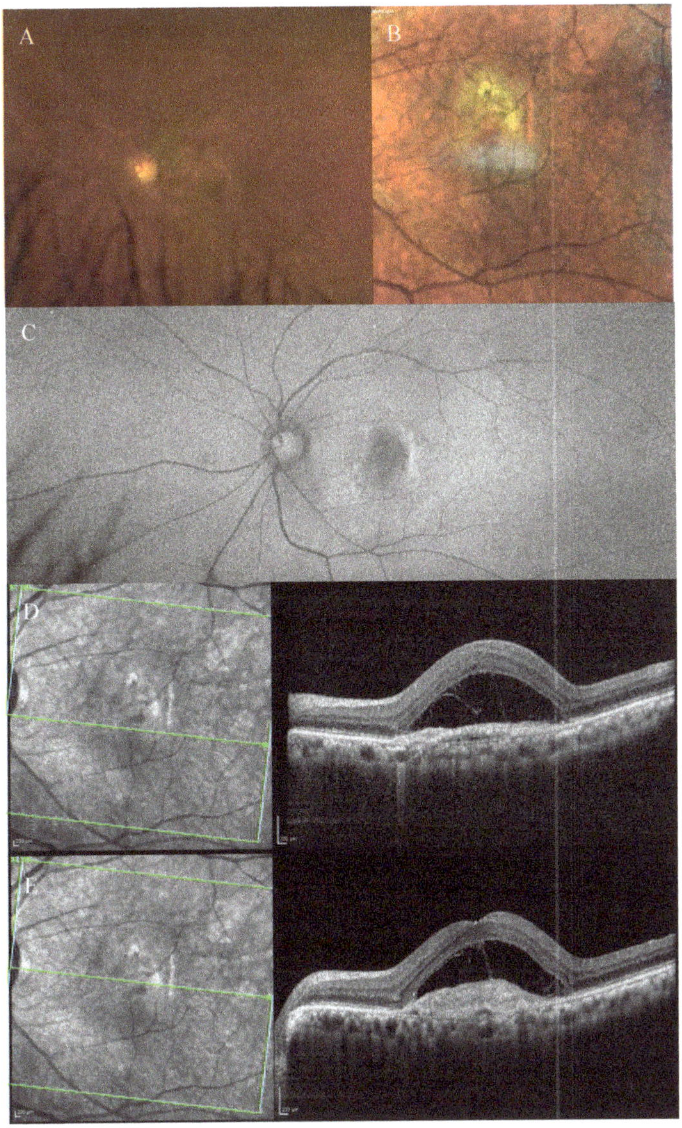

Figure 4. Case 3. Multimodal imaging features of bacillary layer detachment (BALAD) in a patient with left eye neovascular age-related macular degeneration. (**A**) Optos Ultra-widefield colour shows a macular subretinal fluid with a round yellow border. (**B**) Spectralis MultiColor shows a yellowish lesion with subretinal fibrosis. (**C**) Optos Ultra-widefield autofluorescence imaging shows a macular hypoautofluorescence lesion. (**D**,**E**) Heidelberg Spectralis optical coherence tomography shows a type 1 MNV with fibrovascular pigment epithelial detachment, subretinal fluid, and a split at the myoid zone creating an intraretinal space with multiple hyper-reflective foci and a septa-like arrangement and a hyper-reflective granular band at the anterior border of BALAD. The posterior border showed a hyper-reflective thickened band. The external limiting membrane is recognizable anterior to the BALAD.

3.4. Case 4

An 82-year-old male patient presented to our care in June 2019 with a drop of vision in the right eye. His BCVA was 0.7 and 0.3 logMAR in the right and left eye, respectively.

The anterior segment examination was unremarkable in both eyes. An ophthalmological assessment showed features of right nAMD with BALAD and left vitreomacular traction.

Fundus examination showed a right macular yellowish slightly elevated lesion with subretinal haemorrhage (Figure 5A). FAF revealed a hypoautofluorescent macular lesion (Figure 5B).

The OCT scan showed a type 1 MNV with subretinal and intraretinal fluid and BALAD in the right eye. The BALAD cavity presented hyper-reflective granular foci (Figure 5C).

The left eye OCT scan showed vitreomacular traction.

The OCTA scan confirmed the presence of the right MNV.

An anti-VEGF intravitreal injection T&E regimen was commenced.

In October 2021, the patient could not continue the treatment due to hospitalization related to other health issues.

In April 2022, he presented to our care complaining of deterioration of vision in his right eye.

The BCVA was CF, and the anterior segment examination was unremarkable. The OCT scan showed worsening of the BALAD and nAMD features. Thus, the anti-VEGF intravitreal injections treatment was recommenced.

To date, the OCT scan showed an improvement in the subretinal and intraretinal fluid and a partial resolution of the BALAD. Hyper-reflective foci were present in the BALAD cavitation (Figure 5D).

However, no visual gain was observed.

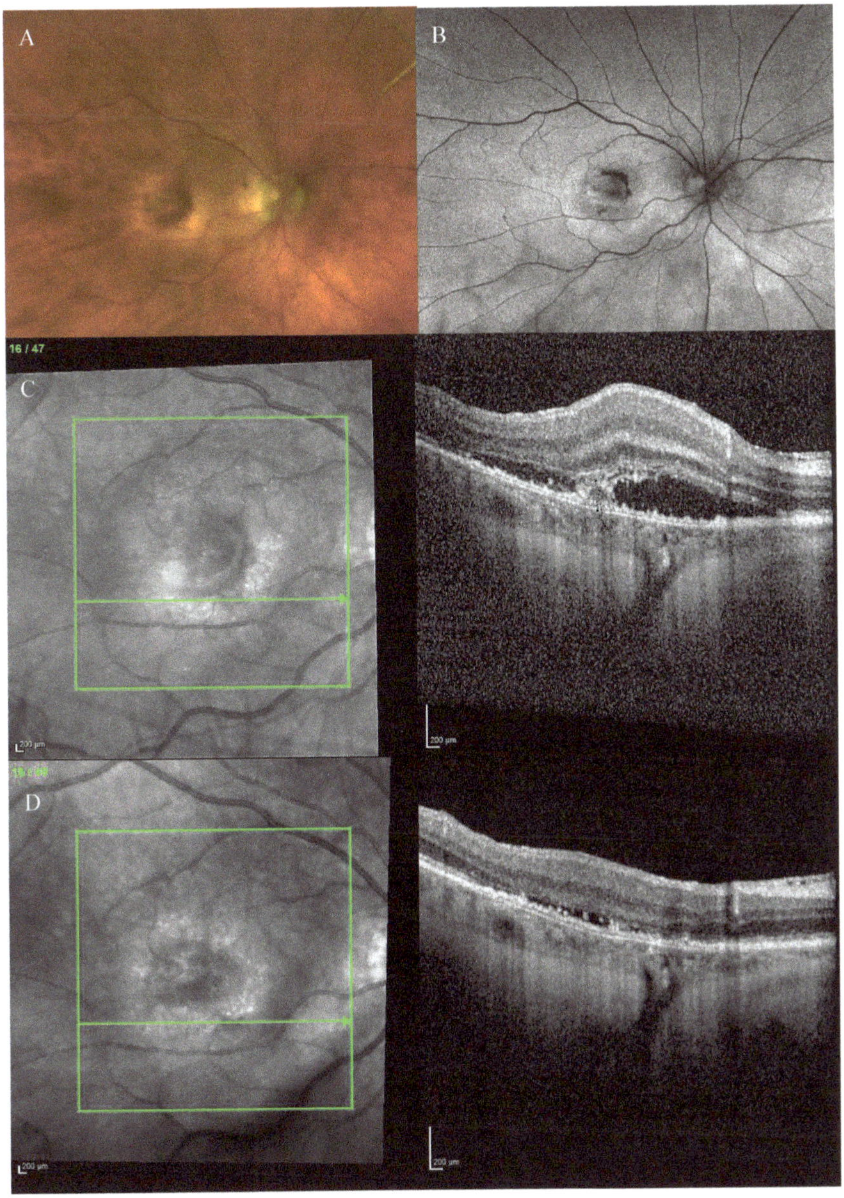

Figure 5. Case 4. Multimodal imaging features of bacillary layer detachment (BALAD) in a right eye neovascular age-related macular degeneration patient. (**A**) Optos Ultra-widefield colour shows a macular yellowish slightly elevated lesion with retinal haemorrhage. (**B**) Optos Ultra-widefield autofluorescence imaging shows a hypoautofluorescent macular lesion. (**C**) Heidelberg Spectralis optical coherence tomography shows a type 1 MNV with subretinal and intraretinal fluid and BALAD. (**D**) Heidelberg Spectralis optical coherence tomography shows a partial resolution of BALAD after anti-vascular endothelial growth factor intravitreal injections. Hyper-reflective foci are present in the BALAD cavitation.

3.5. *Case 5*

A 78-year-old female patient was referred to our department with nAMD in both eyes. She had a long history of different types of anti-VEGF intravitreal injections since 2012.

She showed features of BALAD in both eyes since the initial diagnosis was made. Her initial BCVA was 0.4 logMAR in both eyes.

The anterior segment examination was unremarkable in both eyes.

Fundus examination showed a macular yellowish elevated lesion in both eyes (Figure 6A,B). FAF revealed a hypoautofluorescent macular lesion (Figure 6C,D).

The OCT showed a type 1 MNV with a fibrovascular PED, subretinal fluid, and BALAD in both eyes (Figure 6E,F).

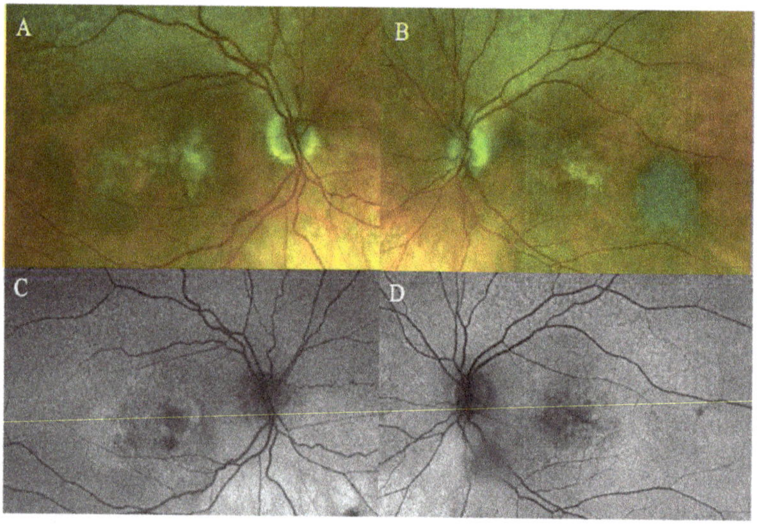

Figure 6. *Cont.*

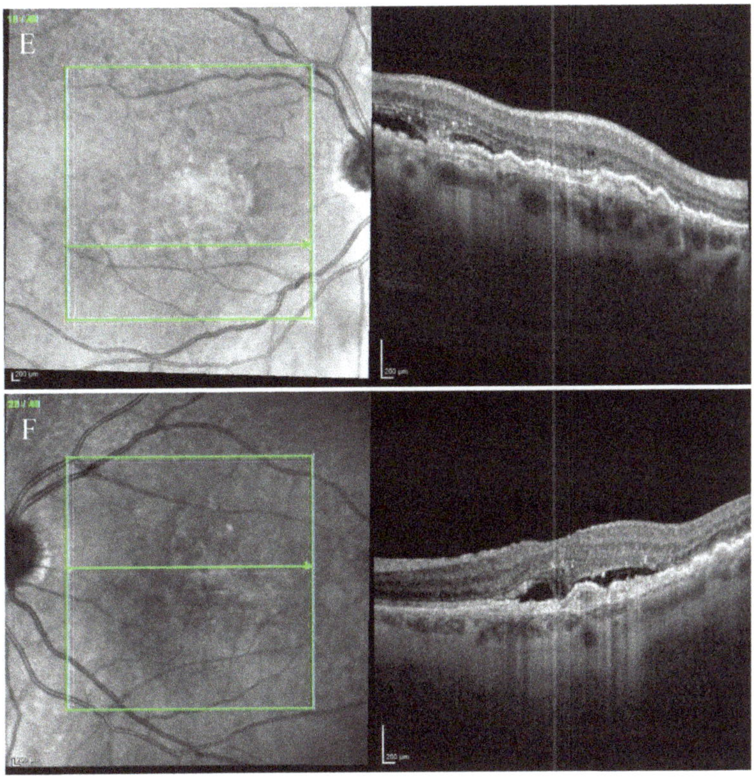

Figure 6. Case 5. Multimodal imaging features of bacillary layer detachment (BALAD) in neovascular age-related macular degeneration patients in both eyes. (**A**,**B**) Optos Ultra-widefield colour shows a macular yellowish slightly elevated macular lesion. (**C**,**D**): Optos Ultra-widefield autofluorescence imaging shows a hypoautofluorescent macular lesion. (**E**) Heidelberg Spectralis optical coherence tomography shows a type 1 MNV with subretinal and intraretinal fluid and BALAD in the right eye. (**F**) Heidelberg Spectralis optical coherence tomography shows a type 1 MNV with subretinal fluid and BALAD in the left eye.

The OCTA scan confirmed the presence of MNV (Figure 7A,B).

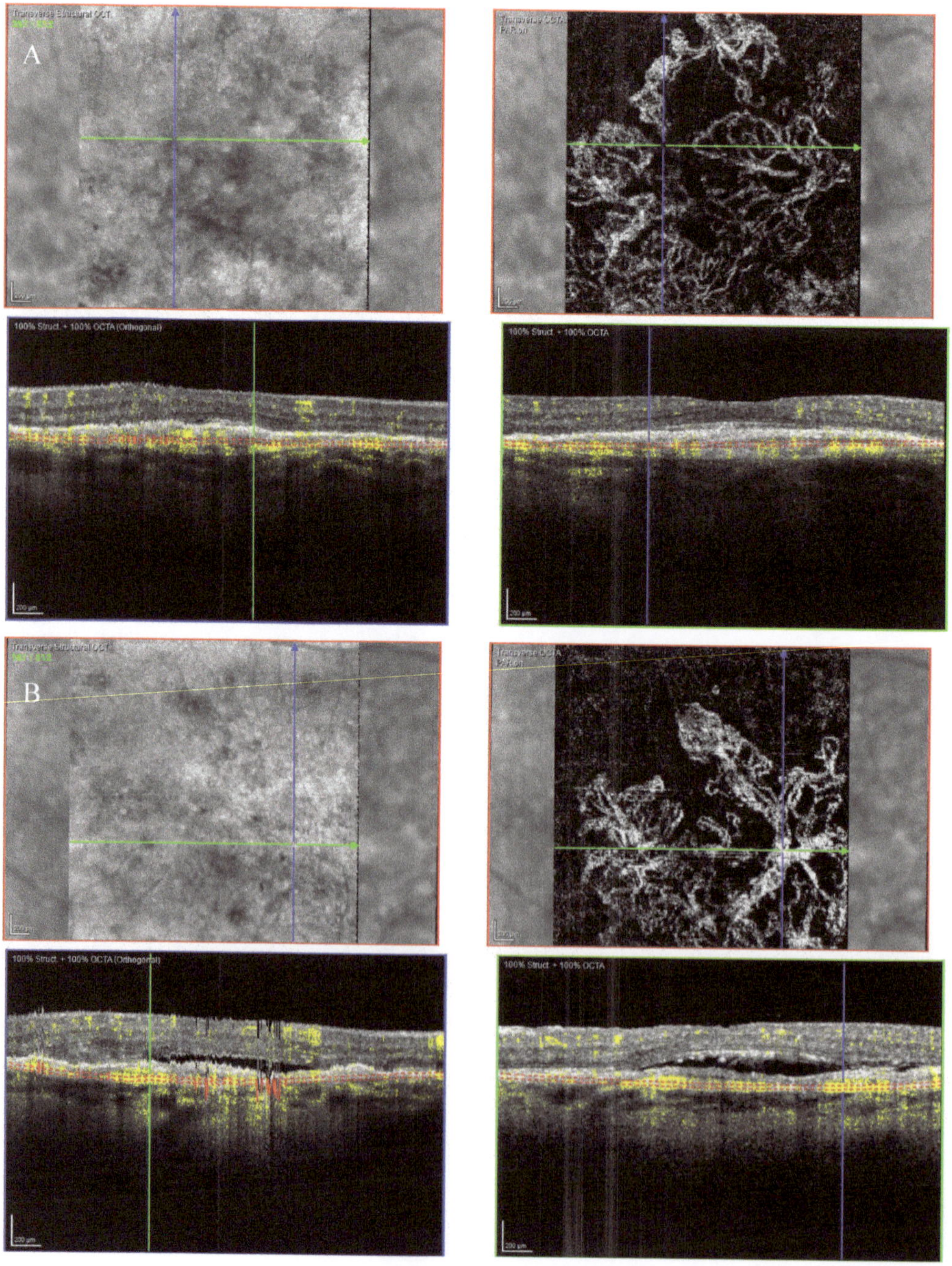

Figure 7. Case 5. Heidelberg Spectralis optical coherence tomography angiography shows a subretinal neovascular membrane in right (**A**) and left eyes (**B**).

Over the years, these features have shown improvement phases alternating to worsening ones.

The patient is still on treatment with anti-VEGF intravitreal injections (T&E regimen) for both eyes.

The BCVA has been stable throughout the treatment period. When last examined, it was 0.63 logMAR in the right eye and 0.32 logMAR in the left eye.

3.6. *Case 6*

A 79-year-old male patient presented to our facility in February 2019 complaining of vision distortion in the left eye. His BCVA was 0.2 logMAR in both eyes.

The anterior segment examination was unremarkable. An ophthalmological assessment showed features of right dry AMD and left nAMD with BALAD.

Fundus examination showed right macular drusen and a left macular yellowish elevated lesion with massive retinal haemorrhage.

FAF revealed a hypoautofluorescent macular lesion in the left eye (Figure 8).

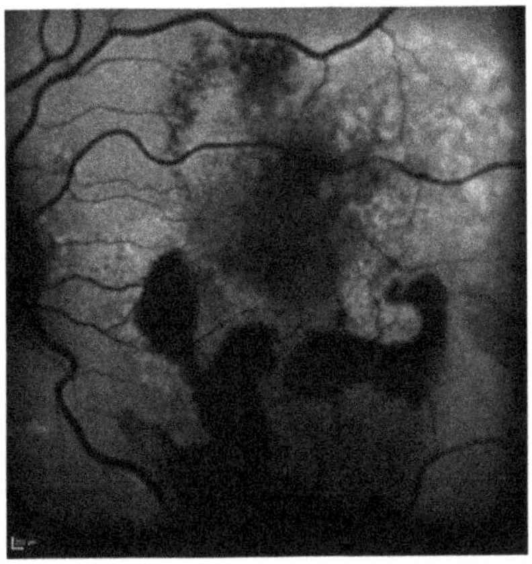

Figure 8. Case 6. Heidelberg Spectralis blue autofluorescence shows a left eye macular hypoautofluorescent lesion.

The OCT scan showed a dry macula with drusen in the right eye and a type 1 MNV with PEDs, intraretinal and subretinal fluid, subretinal haemorrhage, and a haemorrhagic BALAD in the left eye (Figure 9A).

The OCTA scan confirmed the presence of the left MNV.

A T&E regimen with anti-VEGF intravitreal injection therapy was commenced. The patient showed an initial good response to the treatment (Figure 9B). Unfortunately, over the following years, the worsening of the condition was observed despite treatment. Thus, the patient switched to a different anti-VEGF molecule.

To date, the patient is still on treatment. The left BCVA was 0.4 logMAR when last examined. The last OCT scan showed intraretinal fluid and an improved BALAD (Figure 9C).

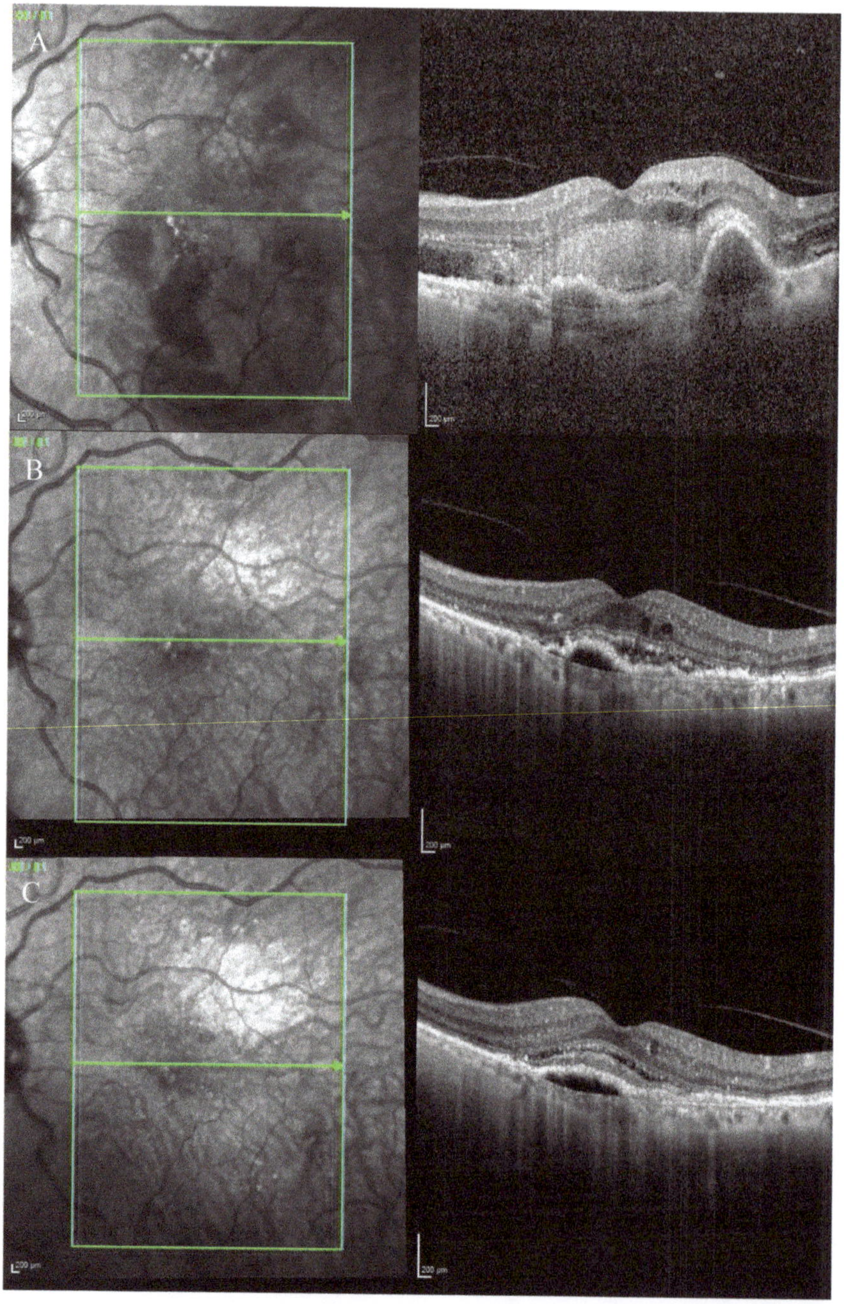

Figure 9. Case 6. (**A**) Heidelberg Spectralis optical coherence tomography shows a type 1 MNV with fibrovascular pigment epithelial detachment, subretinal and intraretinal fluid, and haemorrhagic BALAD in the left eye. (**B**,**C**) Heidelberg Spectralis optical coherence tomography shows a good response after a year and two years of intravitreal anti-VEGF treatment, respectively.

Figure 10 shows Optos Ultra-widefield colour and autofluorescence imaging of the left eye macular lesion two years after the start of the treatment.

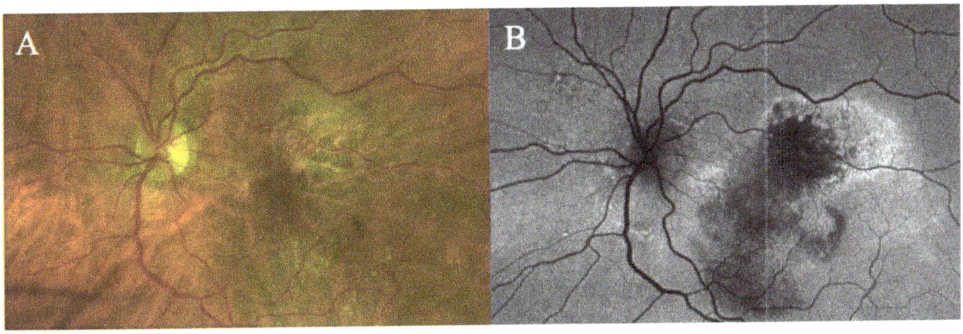

Figure 10. Case 6. Optos Ultra-widefield colour and autofluorescence imaging of the left eye after two years of intravitreal anti-VEGF treatment, showing a yellowish (**A**) and hypofluorescent (**B**) macular lesion. The retinal haemorrhage was fully resolved (**A**,**B**).

3.7. Case 7

A 93-year-old male patient with a diagnosis of peripapillary choroidal neovascularization in the right eye, for which he has been receiving intravitreal injections for over 7 years, presented to our facility.

His BCVA was 0.5 and 0.1 logMAR in the right and left eye, respectively. The anterior segment examination was unremarkable in both eyes.

Fundus examination revealed a right greyish peripapillary lesion and macular drusen in the left eye. FAF was also performed, showing a right peripapillary hypoautofluorescent area (Figure 11).

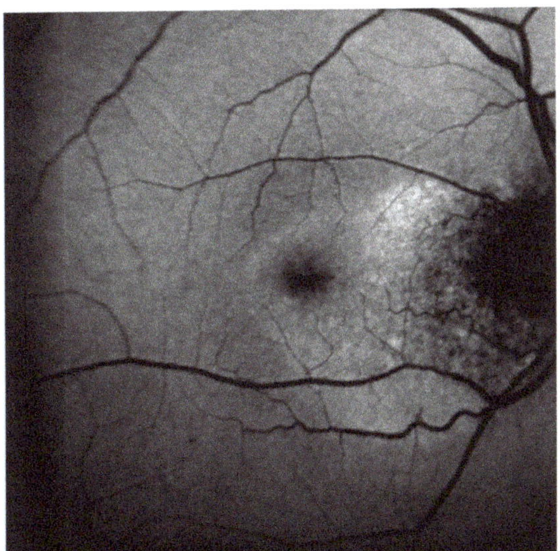

Figure 11. Case 7. Heidelberg Spectralis blue autofluorescence shows a peripapillary hypoautofluorescent lesion in the right eye.

The OCT scan demonstrated a type 1 MNV with PED and BALAD in the right eye. The ceiling of the BALAD presented hyper-reflective foci. The EZ zone was attenuated at the floor (Figure 12).

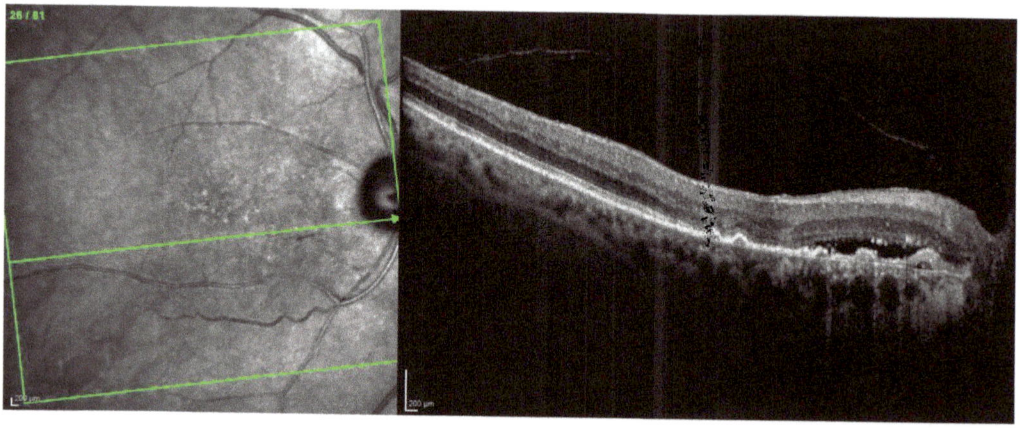

Figure 12. Case 7. Heidelberg Spectralis optical coherence tomography shows a peripapillary type 1 MNV with subretinal fluid and BALAD in the right eye.

The OCTA scan confirmed the presence of the right peripapillary MNV (Figure 13).

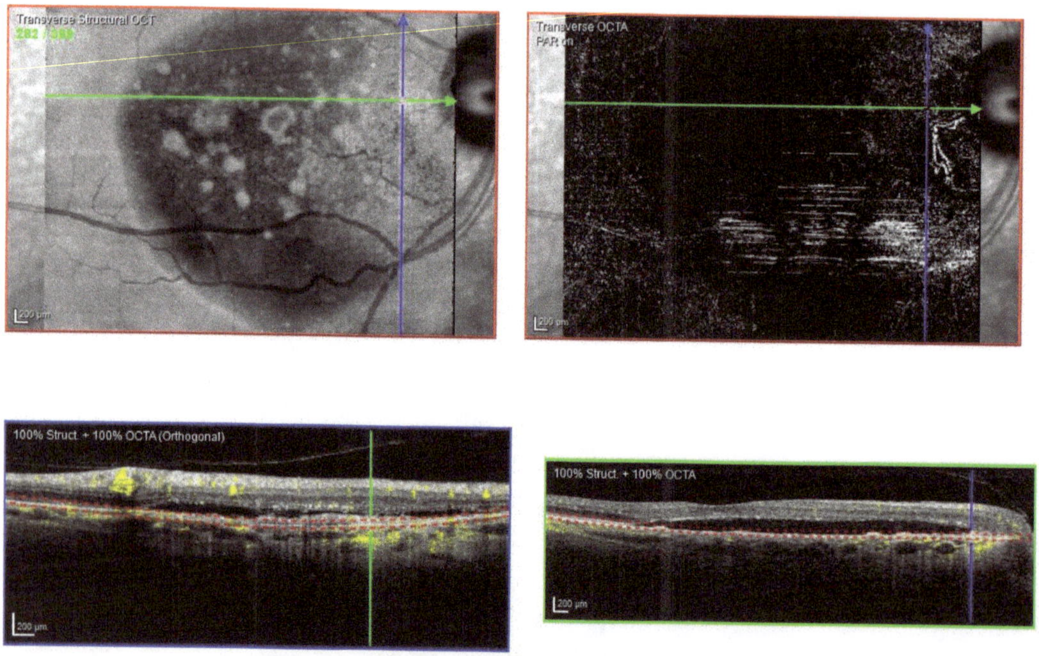

Figure 13. Case 7. Heidelberg Spectralis optical coherence tomography angiography shows a peripapillary subretinal neovascular membrane in the right eye.

These features were shown to be partially responsive to the treatment. The BVCA was stable over the treatment period.

4. Discussion

The term BALAD was recently introduced to indicate a splitting of the photoreceptor layer at the myoid level [1]. Assuming that the photoreceptors' IS myoid is weaker than the ellipsoid and the junctional complexes of the ELM, it has been postulated that the main driving factor for developing BALAD could be the intrinsic weakness of the photoreceptors' IS myoid [1,21,22]. When chorioretinal inflammation and exudation occur, the forces that promote the attachment of the photoreceptors' outer segment (OS) to the RPE may be superior to the tensile strength of the photoreceptors' IS myoid, resulting in splitting [1,21,22]. In nAMD, an IS photoreceptor shedding occurs due to degeneration. It may predispose nAMD patients to develop BALAD [23,24].

We examined the clinical and multimodal imaging findings of eight eyes of seven patients with neovascular age-related macular degeneration who developed bacillary layer detachment.

As reported in the literature, we had non-haemorrhagic BALAD in seven eyes and one haemorrhagic BALAD in one eye.

Ramtohul et al. found that there was a twofold risk of the development of fibrosis through 4 years in eyes presenting with haemorrhagic BALAD [12,20]. The haemorrhagic BALAD was reported in a case of MNV associated with macular telangiectasia type 2 [12]. It is important to differentiate between haemorrhagic BALAD and a submacular haemorrhage because the latter may require additional surgical treatment. As known in the literature, the resolution of macular haemorrhage in nAMD patients may result in a gradual transformation of the sub-RPE fluid into subretinal fibrosis and severe photoreceptor loss [20].

In our report, haemorrhagic BALAD showed a hyper-reflective cavity. These findings made the recognition of the retinal layers difficult. The patient showed an initial good anatomical response, but, over the following years, worsening of the condition was observed. The patient's final BCVA was worse than the initial presentation due to the development of photoreceptor loss and subretinal fibrosis.

The non-haemorrhagic BALAD cavity showed hypo- or hyper-reflective material. In some cases, the hyper-reflective material may have a septa configuration. Some of our patients had a septa configuration of non-haemorrhagic BALAD. It is particularly evident in case reports 1 and 3.

It has been hypothesized that this material in the BALAD cavity consists of inflammatory elements, such as fibrin, and photoreceptor debris [1,21].

We postulated that some NVMs have an intense exudative and inflammatory component with consequent strong traction in the photoreceptor layer and formation of BALAD. It could be sustained by the inflammatory findings in the BALAD cavity.

We reported hyper-reflective granular foci at the ceiling of the BALAD. As reported in the literature, they could represent the myoid zone fragments and photoreceptors IS and OS. The ELM was visible superior to the BALAD on OCT scans.

We also observed a hyper-reflective band contiguous to an attenuated EZ on the floor of the BALAD. These findings have been reported in the literature [10,14–16,20,21]. Ramtohul et al. postulated that floor hyper-reflectivity could be related to the remaining mitochondria of the residual photoreceptor IS on the RPE–basal lamina–Bruch membrane complex. Alternatively, they hypothesized that the BALAD floor signal may be attenuated by the overlying fluid, the myoid fragments, and the OS attached to the RPE–basal lamina–Bruch membrane complex [21]. To date, a few studies have reported BALAD in neovascular AMD (nAMD). Thus, the incidence of BALAD in nAMD is not well defined. Sari Yordi et al. noticed that the bacillary detachment was present in 7.4% (6 of 81) of the eyes reviewed. It was predominantly associated with type 1 and 2 mixed MNV, higher fluid volumes, increased EZ attenuation, and sub-RPE disease [15]. Instead, Jae Hui Kim et al. reported an incidence of bacillary detachment in 4.5% (20 of 442) of the patients examined [16]. In a Korean cohort, Kim et al. observed that the incidence of BALAD was significantly different in the types of MNV. Type 2 MNV was found to be the most associated with

BALAD [16]. These results accorded to the findings of Jung et al. [10]. Conversely, in the study by Ramtohul et al., type 1 MNV was reported to be the most associated with BALAD. They demonstrated BALAD in association with other MNV subtypes [20]. In our study, all cases presented a type 1 MNV, except for one patient who presented a type 2 MNV. Thus, our study is in agreement with the results reported by Ramtohul et al. The role of the presence of BALAD in the progression of nAMD is uncertain.

Ramtohul et al. hypothesized that the typical presence of extracellular matrix proteins in the subretinal space and the intra- and/or subretinal haemorrhages in BALAD may increase the risk of developing macular fibrosis in nAMD, after the fluid resorption. They found a similar distribution of subretinal fibrosis among the MNV subtypes, anti-VEGF agents, and treatment protocols [20]. We observed the development of macular fibrosis in two cases.

However, since anti-VEGF therapy in itself creates a profibrotic environment, at the moment it is not possible to define the role of BALAD in the development of fibrosis. Further studies are necessary.

In one of our cases, we noticed a spontaneous retinal pigment epithelial (RPE) tear in a type 2 MNV after four anti-VEGF intravitreal injections. To the best of our knowledge, there is no evidence of the relationship between BALAD and RPE rip development in the literature.

RPE tears have been reported as part of the natural history of nAMD. Risk factors in patients with AMD receiving anti-VEGF medications include larger lesion linear diameter, vertical PED dimension, and duration of PED formation. Bird proposed that RPE tears are a result of hydrostatic pressure from fluid or material accumulating in the sub-RPE space (in the PED) and eventually causing the rupture of RPE. However, there is a low incidence of retinal pigment epithelium tears after anti-VEGF intravitreal injections [25–27].

Mehta et al. suggested that a hydrostatic force from the choroid strong enough to split the photoreceptors was needed to form a BALAD [1].

Based on the probable assumptions of the BALAD formation and RPE tear development, we hypothesize that there may be some mutual mechanisms in the development of both. BALAD could be an additional risk factor for the early development of an RPE rip, but studies with longer follow-ups of these patients are needed to confirm this hypothesis.

Case series reported in the literature assumed that COVID-19 infection and/or vaccination could be a risk factor in developing BALAD, considering the increased systemic inflammatory response in some patients and infection/vaccination with possible signs of subclinical ocular inflammation. Raphaela M. Fuganti et al. speculated that the occurrence of BALAD and a large amount of fibrin accretion in their patient with central serous chorioretinopathy (CSC) could be related to the systemic inflammatory condition observed in patients with COVID-19 [11]. However, this additional risk factor could not be identified in our study.

Currently, it remains unclear whether BALAD increases the risk of macular atrophy and fibrosis in long-term follow-up. Furthermore, it is unknown if the patients with BALAD need a specific follow-up or a specific treatment regimen. The current treatment of nAMD with BALAD remains the anti-VEGF therapy. Despite this, new treatment modalities are being developed for the treatment of retinal pathologies. More recently, to overcome the drawbacks associated with injecting anti-VEGF, researchers have developed novel ocular pharmacological nanoformulations with multiple bioactive properties for enhanced treatment of age-related macular degeneration [28]. In conclusion, more studies with longer follow-ups are necessary for better defining physiopathology, epidemiology, and the actual role of the presence of BALAD in nAMD.

Author Contributions: Conceptualization, L.F., F.P. and S.Y.; Methodology, F.P. and L.F.; Software, F.P., L.F. and W.R.; Validation, S.Y.; Formal analysis, F.P. and L.F. and S.Y.; Investigation, F.P., L.F. and W.R.; Resources, S.Y.; Data curation L.F., F.P. and W.R.; Writing—original draft preparation, F.P. and L.F.; Writing—review and editing, F.P., L.F. and W.R.; Visualization, S.Y.; Supervision, S.Y.; Project administration, S.Y. All authors have read and agreed to the published version of the manuscript.

Funding: This research received no external funding.

Institutional Review Board Statement: This case series complied with the guidelines of the Declaration of Helsinki. The study was a retrospective analysis of data obtained in clinical routine care collected from the electronic medical record. All patients were treated "on-label". No ethics committee approval was needed for the retrospective review of the anonymized data. No patient-identifiable data was used, and the patients were not identifiable ti the research team in carrying out the research.

Informed Consent Statement: Informed consent was obtained from all subjects involved in the study.

Data Availability Statement: The data are not publicly available due to confidentially issues and will only be available on reasonable request from the corresponding authors.

Conflicts of Interest: The authors declare no conflict of interest.

References

1. Mehta, N.; Chong, J.; Tsui, E.; Duncan, J.L.; Curcio, C.A.; Freund, K.B.; Modi, Y. PRESUMED FOVEAL BACILLARY LAYER DETACHMENT IN A PATIENT WITH TOXOPLASMOSIS CHORIORETINITIS AND PACHYCHOROID DISEASE. *Retin. Cases Brief Rep.* **2021**, *15*, 391–398. [CrossRef]
2. Fernández-Vigo, J.I.; Moreno-Morillo, F.J.; Valverde-Megías, A.; Burgos-Blasco, B.; López-Guajardo, L.; Donate-López, J. ACUTE EXUDATIVE MACULOPATHY AND BACILLARY LAYER DETACHMENT IN PATIENTS WITH CENTRAL SEROUS CHORIORETINOPATHY AFTER PHOTODYNAMIC THERAPY. *Retina* **2022**, *42*, 859–866. [CrossRef] [PubMed]
3. Ledesma-Gil, G.; Desmettre, T.; Mainster, M.A. Bacillary Layer Detachment after Photodynamic Therapy for Central Serous Chorioretinopathy. *Retin. Cases Brief Rep.* **2021**. [CrossRef]
4. Murillo, S.A.; Medina, S.P.; Romero, R.M.; Murillo, F.H. Bacillary Layer Detachment in an Atypical Case of Central Serous Chorioretinopathy Associated with High Hyperopia. *Case Rep. Ophthalmol.* **2022**, *13*, 504–510. [CrossRef]
5. Kohli, G.M.; Bhatia, P.; Shenoy, P.; Sen, A.; Gupta, A. Bacillary Layer Detachment in Hyper-acute Stage of Acute Posterior Multifocal Placoid Pigment Epitheliopathy: A Case Series. *Ocul. Immunol. Inflamm.* **2022**, *30*, 703–706. [CrossRef]
6. Ramtohul, P.; Comet, A.; Denis, D. Bacillary Layer Detachment in Peripapillary Pachychoroid Syndrome. *Ophthalmol. Retin.* **2020**, *4*, 587. [CrossRef]
7. Agarwal, A.; Markan, A.; Aggarwal, K.; Gupta, V. Bacillary layer detachment in tubercular choroidal granuloma: A new optical coherence tomography finding. *Indian J. Ophthalmol.* **2020**, *68*, 1944–1946. [CrossRef]
8. Venkatesh, R.; Reddy, N.G.; Pulipaka, R.S.; Mahendradas, P.; Yadav, N.K.; Jayadev, C. Bacillary Layer Detachment in Unilateral Acute Idiopathic Maculopathy: A Report of 2 Cases. *Ocul. Immunol. Inflamm.* **2023**, *31*, 3–6. [CrossRef] [PubMed]
9. Tekin, K.; Teke, M.Y. Bacillary layer detachment: A novel optical coherence tomography finding as part of blunt eye trauma. *Clin. Exp. Optom.* **2019**, *102*, 343–344. [CrossRef] [PubMed]
10. Jung, J.J.; Soh, Y.Q.; Yu, D.J.G.; Rofagha, S.M.; Lee, S.S.; Freund, K.B.; Hoang, Q.V.M. BACILLARY LAYER DETACHMENT BECAUSE OF MACULAR NEOVASCULARIZATION. *Retina* **2021**, *41*, 2106–2114. [CrossRef]
11. Fuganti, R.M.; Casella, A.M.; Roisman, L.; Zett, C.; Maia, M.; Farah, M.E.; Lima, L.H. Case series bacillary layer detachment associated with acute central serous chorioretinopathy in patients with COVID-19. *Am. J. Ophthalmol. Case Rep.* **2022**, *28*, 101690. [CrossRef] [PubMed]
12. Ramtohul, P.; Comet, A.; Denis, D.; Gascon, P. Hemorrhagic Bacillary Layer Detachment in Macular Telangiectasia Type 2. *Retina* **2021**, *41*, e42–e43. [CrossRef] [PubMed]
13. Atas, F.; Kaya, M.; Saatci, A.O. The effect of pulse steroid treatment of ten days' long on the improvement of bacillary layer detachment in a patient with Vogt-Koyanagi Harada disease. *Rom. J. Ophthalmol.* **2021**, *65*, 183–186. [CrossRef] [PubMed]
14. Ramtohul, P.; Denis, D.; Gascon, P. Bacillary Layer Detachment in Acute Posterior Multifocal Placoid Pigment Epitheliopathy: A Multimodal Imaging Analysis. *Retina* **2021**, *41*, e12–e14. [CrossRef] [PubMed]
15. Yordi, S.; Sarici, K.; Cetin, H.; Lunasco, L.M.; Le, T.K.; Sevgi, D.D.; PharmD, R.Z.; Meng, X.; Reese, J.L.; Srivastava, S.K. Bacillary Detachment in Neovascular Age-Related Macular Degeneration: Incidence, Clinical Features, and Response to Anti-VEGF Therapy. *Ophthalmol. Retin.* **2022**, *6*, 1061–1069. [CrossRef]
16. Kim, J.H.; Kim, J.W.; Kim, C.G. BACILLARY LAYER DETACHMENT IN A KOREAN COHORT WITH NEOVASCULAR AGE-RELATED MACULAR DEGENERATION. *Retina* **2022**, *42*, 1028–1037. [CrossRef]
17. Rodríguez-Vidal, C.; Galletero Pandelo, L.; Artaraz, J.; Pereira, A.; Yadav, N.K.; Chhablani, J. Bacillary layer detachment in a patient with serpiginoid choroiditis. *Indian J. Ophthalmol.* **2022**, *70*, 2687–2689. [CrossRef]
18. Venkatesh, R.; Agrawal, S.; Reddy, N.G.; Pereira, A.; Yadav, N.K.; Chhablani, J. Bacillary layer detachment in acute nonpenetrating ocular trauma. *Can. J. Ophthalmol.* **2022**, *57*, 328–336. [CrossRef]
19. Guner, M.K.; Starr, M.R. Bacillary Layer Detachment in Age-Related Macular Degeneration. *Ophthalmol. Retin.* **2022**, *6*, 722. [CrossRef]
20. Ramtohul, P.; Malclès, A.; Gigon, E.; Freund, K.B.; Introini, U.; Bandello, F.; Cicinelli, M.V. Long-Term Outcomes of Bacillary Layer Detachment in Neovascular Age-Related Macular Degeneration. *Ophthalmol. Retin.* **2022**, *6*, 185–195. [CrossRef]

21. Ramtohul, P.; Engelbert, M.; Malclès, A.; Edward, G.; Elisabetta, M.; Giulio, M.; Eduardo, C.d.S.; Cagri, B.; Curcio, C.A.; Bailey, F.K. BACILLARY LAYER DETACHMENT: MULTIMODAL IMAGING AND HISTOLOGIC EVIDENCE OF A NOVEL OPTICAL COHERENCE TOMOGRAPHY TERMINOLOGY: Literature Review and Proposed Theory. *Retina* **2021**, *41*, 2193–2207. [CrossRef] [PubMed]
22. Liakopoulos, S.; Keane, P.A.; Ristau, T.; Kirchhof, B.; Walsh, A.C.; Sadda, S.R. Atypical outer retinal fluid accumulation in choroidal neovascularization: A novel OCT finding. *Ophthalmic Surg Lasers Imaging Retin.* **2013**, *44* (Suppl. 6), S11–S18. [CrossRef]
23. Litts, K.M.; Messinger, J.D.; Freund, K.B.; Zhang, Y.; Curcio, C.A. Inner Segment Remodeling and Mitochondrial Translocation in Cone Photoreceptors in Age-Related Macular Degeneration With Outer Retinal Tubulation. *Investig. Ophthalmol. Vis. Sci.* **2015**, *56*, 2243–2253. [CrossRef]
24. Litts, K.M.; Zhang, Y.; Freund, K.B.; Curcio, C.A. OPTICAL COHERENCE TOMOGRAPHY AND HISTOLOGY OF AGE-RELATED MACULAR DEGENERATION SUPPORT MITOCHONDRIA AS REFLECTIVITY SOURCES. *Retina* **2018**, *38*, 445–461. [CrossRef]
25. Bird, A.C. Doyne Lecture. Pathogenesis of retinal pigment epithelial detachment in the elderly; the relevance of Bruch's membrane change. *Eye* **1991**, *5 Pt 1*, 1–12. [CrossRef] [PubMed]
26. Ersoz, M.G.; Karacorlu, M.; Arf, S.; Sayman Muslubas, I.; Hocaoglu, M. Retinal pigment epithelium tears: Classification, pathogenesis, predictors, and management. *Surv. Ophthalmol.* **2017**, *62*, 493–505. [CrossRef] [PubMed]
27. Vazquez-Alfageme, C.; Nicholson, L.; Hamilton, R.D.; Patel, P.J. INCIDENCE AND LONG-TERM VISUAL ACUITY OUTCOMES OF RETINAL PIGMENT EPITHELIUM TEARS AFTER INTRAVITREAL ANTI-VASCULAR ENDOTHELIAL GROWTH FACTOR TREATMENT OF NEOVASCULAR AGE-RELATED MACULAR DEGENERATION. *Retina* **2019**, *39*, 664–669. [CrossRef] [PubMed]
28. Nguyen, D.D.; Luo, L.J.; Yang, C.J.; Lai, J.Y. Highly Retina-Permeating and Long-Acting Resveratrol/Metformin Nanotherapeutics for Enhanced Treatment of Macular Degeneration. *ACS Nano* **2023**, *17*, 168–183. [CrossRef]

biomedicines

Article

Advanced ImageJ Analysis in Degenerative Acquired Vitelliform Lesions Using Techniques Based on Optical Coherence Tomography

Ioana Damian [1,*], George-Adrian Muntean [1], Larisa-Bianca Galea-Holhoș [2] and Simona-Delia Nicoară [1,3]

1 Department of Ophthalmology, "Iuliu Hatieganu" University of Medicine and Pharmacy, 8 Victor Babeș Street, 400012 Cluj-Napoca, Romania; georgemuntean99@gmail.com (G.-A.M.); stalu@umfcluj.ro (S.-D.N.)
2 Department of Anatomy, Faculty of Medicine and Pharmacy, University of Oradea, 1 Decembrie Street, 410068 Oradea, Romania; larihholhos@gmail.com
3 Clinic of Ophthalmology, Emergency County Hospital, 3–5 Clinicilor Street, 400006 Cluj-Napoca, Romania
* Correspondence: ioana.damian@umfcluj.ro

Abstract: Acquired vitelliform lesions (AVLs) are associated with a large spectrum of retinal diseases, among which is age-related macular degeneration (AMD). The purpose of this study was to characterize AVLs' evolution in AMD patients using optical coherence tomography (OCT) technology and ImageJ software. We measured AVLs' size and density and followed their impacts over surrounding retinal layers. Average retinal pigment epithelium (RPE) thickness in the central 1 mm quadrant (45.89 ± 27.84 μm vs. 15.57 ± 1.40 μm) was significantly increased, as opposed to the outer nuclear layer (ONL) thickness, which was decreased (77.94 ± 18.30 μm vs. 88.64 ± 7.65 μm) in the vitelliform group compared to the control group. We found a continuous external limiting membrane (ELM) in 55.5% of the eyes compared to a continuous ellipsoid zone (EZ) in 22.2% of the eyes in the vitelliform group. The difference between the mean AVLs' volume at baseline compared to the last visit for the nine eyes with ophthalmologic follow-up was not statistically significant ($p = 0.725$). The median follow-up duration was 11 months (range 5–56 months). Seven eyes (43.75%) were treated with intravitreal anti-vascular endothelium growth factor (anti-VEGF) agent injections, in which we noted a 6.43 ± 9 letter decrease in the best-corrected visual acuity (BCVA). The increased RPE thickness could suggest hyperplasia contrary to the decreased ONL, which could mirror the impact of the vitelliform lesion on photoreceptors (PR). Eyes that received anti-VEGF injections did not show signs of improvement regarding BCVA.

Keywords: vitelliform lesions; optical coherence tomography; image analysis; ImageJ; anti-VEGF

Citation: Damian, I.; Muntean, G.-A.; Galea-Holhoș, L.-B.; Nicoară, S.-D. Advanced ImageJ Analysis in Degenerative Acquired Vitelliform Lesions Using Techniques Based on Optical Coherence Tomography. *Biomedicines* **2023**, *11*, 1382. https://doi.org/10.3390/biomedicines11051382

Academic Editor: Oyuna Kozhevnikova

Received: 31 March 2023
Revised: 26 April 2023
Accepted: 1 May 2023
Published: 6 May 2023

1. Introduction

The vitelliform macular lesion is classically found in Best's vitelliform dystrophy but similar lesions can be identified in adults as "acquired vitelliform lesions" (AVLs). Gass described in 1974 a "peculiar foveomacular dystrophy", which he further named "foveomacular vitelliform dystrophy: adult type" [1]. Over time, several terms have been applied to disorders similar to Gass' description, such as vitelliform macular lesion, pseudo-vitelliform macular dystrophy, vitelliform macular degeneration, vitelliform lesion in adults, pseudo-vitelliform macular degeneration or adult vitelliform macular detachment in patients with basal laminar drusen [2].

The vitelliform lesion represents an accumulation of lipofuscin, melanosomes, melanolipofuscin and outer segment debris in the subretinal space, between the photoreceptors (PR) layer and retinal pigment epithelium (RPE) [3]. Since AVLs are mainly composed of pigmented granules of RPE origin, it was speculated by some authors that they might denote a specific RPE stress response to an unclarified insult [3]. Other authors observed a lack of direct apposition between the PR outer segments and RPE, which could be responsible for a delayed phagocytosis of shedding the outer segment photoreceptor tips, leading to the accumulation of yellow subretinal vitelliform material [4]. A clinicopathological case series performed on 14 eyes demonstrated two sources for the vitelliform material: an internal one, from the PR outer segment discs and an external one, from the RPE [1]. Electron-microscopy-based studies quantified the deposit's ultrastructure and found that the largest component was a flocculent accumulation of small heterogeneous shapes (>50%), the second component was defined as "other", the third component was RPE granules, the fourth component was smooth spherical profiles with homogeneous interiors resembling lipid droplets and the fifth component was constituted of PR outer-segments-derived materials [5].

Vitelliform lesions encompass a wide clinical spectrum of macular pathology such as degenerative (age-related macular degeneration—AMD), dystrophic (adult-onset foveomacular dystrophy—AOFVD), paraneoplastic, toxic (deferoxamine toxicity) as well as vitreoretinal interface (epiretinal membrane—ERM, vitreomacular traction—VMT) disorders [3,6]. The association between AVLs and cuticular drusen or chronic central serous chorioretinopathy was also noted [6]. Juliano et al. named the degenerative vitelliform lesions present in the elderly acquired vitelliform macular degeneration (AVMD). It occurs in cases without a family history of the disease, compared to an autosomal dominant inheritance pattern that characterizes the dystrophic vitelliform cases [7].

The relationship between AMD and vitelliform lesions was first noted by Gass. When AVL is associated with AMD, beside drusen, patients could also display subretinal drusenoid deposits (SDDs), hyperreflective foci (HRF), subretinal fluid (SRF), pigment epithelial detachments (PEDs), or macular neovascularization (MNV) [5]. The natural course of AVLs could be further complicated by choroidal neovascularization or foveal atrophy [3]. Thus, a study was carried out to analyze AVLs' evolution in AMD eyes and revealed that 12% of AVLs led to type 1 MNV and 44% led to atrophy [5]. Due to the fact that there seems to be a strong preference of AVL for sub foveal location, it has been suggested that there might be a relationship with cones and their supporting cells [5]. In AMD, similarly to Best's disease and AVMD, lipofuscin accumulates in the lysosomal compartment of the RPE, which seems to be one of the first and most important steps in the disease etiology [8].

The presence of subretinal fluid (SRF) in AVMD sometimes puzzles the clinician who must decide if the patient has a dry AMD converted to wet AMD or SRF with a vitelliform lesion [7,8]. The accumulation of SRF associated with vitelliform lesions was explained by a mechanical displacement between the RPE and outer retinal layers, which consequently inhibits the RPE from pumping out liquified lipofuscin debris [7].

Clinically, AVL is found as a slightly elevated yellow lesion, measuring approximately one-third to one disc diameter (DD) in size, with pigmentations in the form of a spot, figure or ring [1]. In optical coherence tomography (OCT), AVLs are characterized as a distinct layer of hyperreflective material between the PR and an irregularly contoured RPE band [5]. Optical coherence tomography angiography (OCTA)-based studies highlighted that the accumulation of the vitelliform material displaces the capillary network, inducing a decreased density of blood vessels at the superficial and deep capillary plexuses as well as the choriocapillaris [9].

BEST1 gene encodes a basolateral membrane protein called bestrophin 1 specific to RPE cells. This protein acts as an anion channel, having an important role in the transepithelial electrical potential and intracellular Ca^{2+} signaling [10]. Although it is specific for Best's disease, a small number of BEST1 variants have been identified in patients with AMD, such as Thr216Ile and Leu567 alterations, which were found in three out of 259 AMD patients' (1.1%) eyes [11]. Another five different alterations were found in five (1.5%) of 321 AMD patients as shown by Lotery A.J. et al. [12]: three patients carried a missense change (Arg105Cys, Glu119Gln, Val275Ile), while two carried a heterozygous nonsense mutation (Lys149X). Given the low frequency of mutation found in AMD patients, the authors stated that these patients could have been misdiagnosed, being in fact BEST1 mutations that mimic AMD [12].

Regarding AVLs' evolution, the following scenario was described: an expansion phase is common to all AVLs, which is followed by either a collapse to complete RPE and outer retinal atrophy (cRORA), stability or resorption [5]. After reaching maximum expansion, the lesions that collapsed were completely resorbed after 24 months. Moreover, stable or resorbing lesions seem to be characterized by slower phases of growth and contraction [5]. Previous authors observed AVL complicated with MNV especially in the collapsing phase in 12% of study eyes [5].

In cases of AVL and SRF, the current literature offers no consensus regarding the benefit of anti-vascular endothelial growth factor (anti-VEGF) injections. In the context of a diagnostic dilemma, some patients receive anti-VEGF injections being mistaken as having neovascular AMD as opposed to AVL. Previous authors suggested two options: either to monitor closely and to treat only if the SRF increases and the patient becomes symptomatic or to give an anti-VEGF injection and determine the response in two weeks [7]. At this point, if the fluid disappears, there is a high chance of choroidal neovascularization (CNV), but if the SRF remains stable or slightly decreased, it can be assumed that the patient has AVMD [7].

The main purpose of the present study was to describe and analyse AVMD in adult patients using quantitative and qualitative features. A secondary purpose was to evaluate AVMD's evolution in time and its response to anti-VEGF therapy.

2. Materials and Methods

2.1. Study Design

We conducted a retrospective, single-center study that was approved by the Ethics Committee belonging to "Iuliu Hațieganu" University of Medicine and Pharmacy (IHUMP), Cluj-Napoca, Romania (no.273/19.09.2022). The study protocol complied with the rules of the Declaration of Helsinki.

2.2. Study Sample

We reviewed the OCT images of all consecutive patients who were diagnosed with macular pathology between 2017 and 2021, in the setting of the Emergency County Hospital, Clinic of Ophthalmology, Cluj-Napoca. The following combination of keywords were used in the search process: macular dystrophy or macular degeneration. The following diagnoses were found: macular atrophy 2 patients, macular degeneration 19 patients, macular dystrophy 3 patients, AMD 702 patients and maculopathy 34 patients. We further selected only patients with vitelliform lesions. The SD-OCT diagnosis of vitelliform material was based on the presence of a homogeneous or non-homogenous hyperreflective material above the RPE band that could not be classified as drusen or pigment epithelial detachment (PED). Fifteen patients met this criterion. The clinical stage of the disease applied on OCT images was determined according to Best's disease classification: a. pre-vitelliform stage; b. vitelliform or egg yolk stage; c. pseudohypopyon stage; d. vitelliruptive or scrambled egg stage; e. atrophic stage, f. cicatricial and choroidal neovascularization [13]. Further on, 13 patients were considered eligible; 2 patients were excluded because of the poor OCT scans. Finally, a total of 18 eyes belonging to 13 patients with adult vitelliform lesions were included in the study. Because of the lack of family history enquiry and genetic testing or electrooculograms, the term "acquired vitelliform lesion" (AVL), was used to describe the patients with AVMD for the entire study [14].

Patients with an ophthalmologic examination in the same setting between January 2017 and September 2019, with no history of AMD, macular atrophy, macular degeneration or maculopathy were selected for the control group. Thus, we set up 2 groups of patients: vitelliform lesion (18 eyes) and control (14 eyes). The algorithm according to which the 2 groups were created is illustrated in Figure 1.

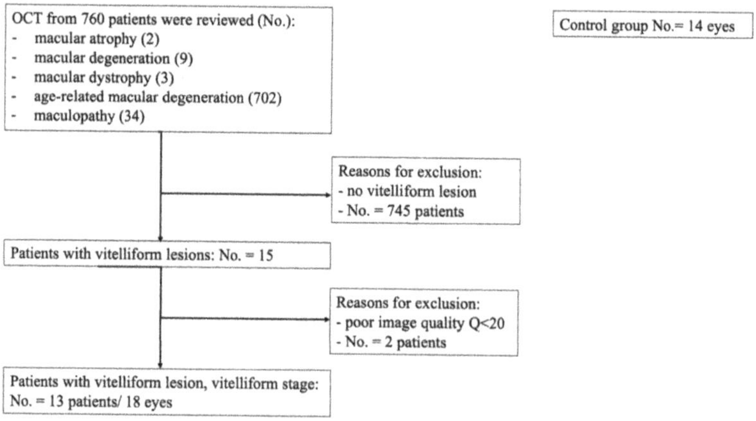

Figure 1. Flow diagram illustrating the study selection process. OCT = Optical coherence tomography; No = number.

Inclusion criteria for the vitelliform group were as follows:

1. Patients with a diagnosis of macular dystrophy or macular degeneration.
2. Minimal criterion for diagnosing AVLs was the presence of a macular round, hyperreflective lesion on the OCT, above the RPE.
3. Only high-quality (Q) OCT scans > 20 db (Q ranging from 0 = poor to 40 = excellent).

Exclusion criteria for vitelliform group:

1. Subretinal fibrosis, CNV, geographic atrophy (GA), epiretinal membrane, macular hole.
2. Eyes with poor-quality images preventing adequate grading Q < 20.

If a patient was confirmed with bilateral vitelliform lesion, both eyes were included in the study.

Participants underwent VA testing with a Snellen visual acuity chart, slit lamp biomicroscopy and dilated eye fundus examination.

The following baseline clinical characteristics were recorded from the patients' data files: age, gender, eye laterality, best-corrected visual acuity (BCVA), follow-up and type of intravitreal injection. Demographic data and ophthalmic examination were collected from the hospital's informatic system. The OCT data were collected from the OCT database.

Patients with at least 2 consecutive examinations and OCT images were further included for a detailed vitelliform lesion analysis, in which we studied the lesion's evolution in time and we measured its volume.

2.3. *Spectral Domain-OCT (SD-OCT) Data Aquisition*

OCT was performed using a Spectralis OCT (Heidelberg Engineering, Inc. Heidelberg, Germany). All frames were reviewed for the presence of AVL by one of the authors (I.D.). The images were obtained after using the automated eye alignment eye tracking software (TruTrack< Heidelberg, Engineering). Two scanning protocols were used: 1. "Fast macular" scanning protocol: 25 horizontal raster lines per eye separated by 240 μm, centered on the fovea, with a 20 × 20° scan and automatic real mean value (ART value) set at 9, was used; or 2. "Posterior pole" scanning protocol: 61 horizontal raster lines per eye separated by 120 μm, centered on the fovea with a 30 × 25° scan and automatic real mean value (ART value) set at 9. Only images with more than a 20 db signal strength and with individual retinal layers that could be identified were used for the analysis. ETDRS 1 mm macular maps were used to report the macular thickness and the central 1 mm ring was defined as the central thickness.

Segmentation was automatically performed using the Spectralis software version 6.0. We defined the following parameters: 1. retinal pigment epithelium layer (RPE)—Bruch's membrane complex—between the outer limit of the photoreceptor layer (PR1/2) and choriocapillaris' inner boundary; 2. outer retina—between ELM and Bruch's membrane but not including the latter; 3. outer nuclear layer (ONL)—between the outer plexiform layer (OPL) and ELM; 4. inner retina from the ILM to the external limiting membrane (ELM); 5. total retina thickness—between the internal limiting membrane (ILM). The inbuilt Spectralis mapping software was used to perform measurements from each SD-OCT scan.

The images were reviewed by one of the authors (I.D.) before data analysis and manual adjustments to the retinal layer segmentation were made if necessary.

In the OCT, the subretinal drusenoid deposit (SDD) was defined as a hyperreflective material above the RPE, between the RPE–Bruch membrane band and the ellipsoid zone (EZ) [5]. Soft drusen was defined as a deposit of material between the RPE–basal lamina (RPE–BL) and the inner collagenous layer of Bruch's membrane [5]. Cuticular drusen was defined as RPE elevations in the form of blunted triangular apices, the bases of which sit on Bruch's membrane with apices towards the neurosensory retina [5]. Subretinal fluid (SRF) appeared as a hyporeflective space located between the external limiting membrane (ELM) and the RPE–BL band [5]. Hyperreflective foci (HRF) were defined as solitary, small (<30 μm) or medium level hyperreflective retinal foci [15].

We described AVL's fate as either: 1. collapsed—meaning a regressed AVL with cRORA, 2. stable AVL—meaning an AVL that did not completely regress during the follow-up, or 3. resorbed AVL, seen as a disappeared lesion with preservation of the surrounding tissue.

2.4. *Vitelliform Lesions Analysis using Fiji Software*

Images were analyzed using the public domain software, Fiji (version 1.53T, https://imagej.nih.gov/ij/ (accessed on 20 March 2022) [16]. Before uploading the OCT images to ImageJ, the vertical-to-horizontal ratio of the image was changed to 1:1 μm. We analyzed the non-stretched OCT scans to overcome the erroneous quantification of parameters [17]. First, the OCT image was opened in Fiji software and the scale was set. We measured a pixel length of 200 μm as given on the scale at the bottom of the OCT scan image, using the line tool. We further used the "magnifying glass" at 400% to obtain a better view of the vitelliform lesion and to easily identify its margins. With the help of the "polygon selection tool", only the region with the vitelliform lesion was selected using the "crop button". As previously defined by other authors, the anterior AVL boundary was considered the external limiting membrane, while the posterior AVL boundary was the retinal pigment epithelium–(basal lamina)–Bruch membrane (RPE–(BL)–BrM(RPE+BL–BrM)) complex [5]. If a PED was found, the posterior boundary was the RPE–BL. For each eye, we analyzed all the scans passing through the AVL and we chose the one passing through the fovea. On this particular scan, using the "straight line" option, we drew the base width and the maximum height of the vitelliform lesion. Using the "freehand line" tool we selected the entire vitelliform lesion to measure its perimeter. We performed 3 measurements for every parameter/eye and averaged them. The parameters are classified as follows:

2.4.1. Vitelliform Lesion Size

a. Maximum height: defined as the distance between the apex of the lesion and the RPE (see Figure 2. A: light green line). We defined apex as the vertex located at the highest point of the AVL or the point furthest from the base.
b. Base width: defined as the distance between either edge of the lesion (see Figure 2A: dark green line).

Using the "freehand line" tool, the vitelliform lesion was delineated; this represents the region of interest (ROI) that was added to the ROI manager. If the AVLs were located on top of a drusenoid PED, we did not include the PED in the measurements. We set the needed measurements such as area and density characteristics, and then we pressed the measure button. We further extracted the results.

c Perimeter: defined as the total length of the vitelliform lesion outside the boundary (see Figure 2B).

d *Area* (*A*) (1): defined as the sum of the areas of each individual pixel, a_p, in calibrated square units, within the borders of the lesion (see Figure 2C):

$$A = \sum a_p, \tag{1}$$

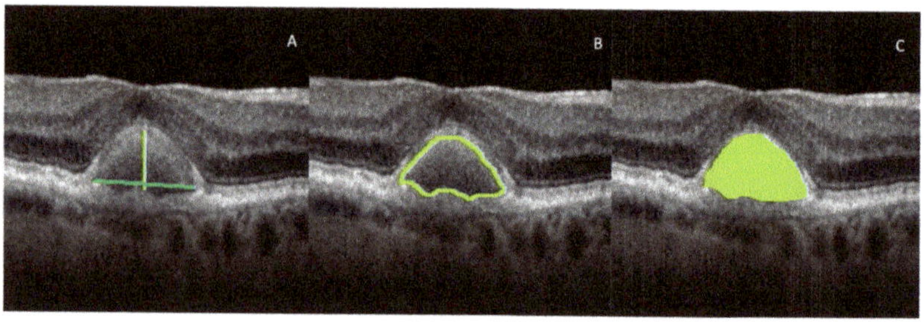

Figure 2. OCT images illustrating vitelliform lesion size parameters. (**A**) Maximum height (green light line) and base width (dark green line); (**B**) perimeter; (**C**) area. The stretched 1 × 1 pixel image was used for illustrative purposes only, to help visualize the details of the retinal structures.

After the vitelliform lesion was delineated with the "freehand line" tool, we further analyzed various vitelliform lesion characteristics, as follows.

2.4.2. Vitelliform Lesion Density

For the density parameters, we used the inbuilt Fiji measurements option.

a. Mean gray value: defined as the sum of the gray values of all the pixels in the selection divided by the number of pixels.
b. Modal gray value: defined as the most frequently occurring gray value within the selection. Corresponds to the highest peak in the histogram. Uses the heading mode.
c. Min gray value: defined as the minimum gray value within the selection.
d. Max gray value: defined as the maximum gray value within the selection.
e. Integrated density (*IntDen*): defined as the product of area and mean gray value (2):

$$\boldsymbol{IntDen = Area \times Mean\ Gray\ value}, \tag{2}$$

f Raw integrated density (RawIntdensity): defined as the sum of the values of the pixels in the vitelliform lesion.

2.4.3. Vitelliform Lesions' Volume and Evolution

For 6 eyes with follow-up OCT images and an OCT Q over 20 dB, we collected all the B-scans that passed through the vitelliform lesions. Afterwards, we uploaded the images, one by one in Fiji, and calculated the area for each one of the segmentations (see Figure 3). In order to estimate the volume of the lesion, we summed up all the area measurements/eye and multiplied it with the distance between the sections (120 or 240 μm).

We compared the lesions' size and density values from the first presentation with the follow-up. We also compared the thickness of the retinal layers from the vitelliform eyes with the normal eyes to find out if there was a connection between the vitelliform lesion's presence and the surrounding layers.

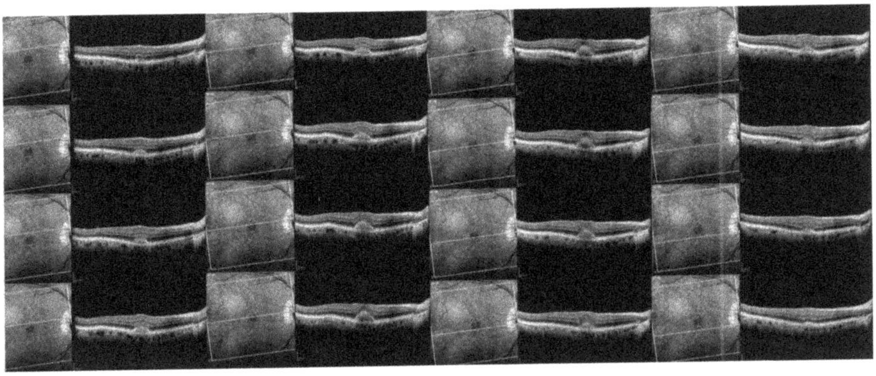

Figure 3. SD-OCT B-scans passing through the entire vitelliform lesion of one clinical study eye.

2.5. Statistical Analysis

The results are presented as percentages (%) for the categorical variables and mean ± standard deviation (SD) for the continuous variables. The normality of the data distribution was confirmed via the Kolmogorov–Smirnov test. We used the parametric Student's *t*-test to compare the variables with a normal distribution between groups and non-parametric Mann–Whitney U tests for variables with a non-normal distribution. We used the *t*-test for the dependent means to compare the values from baseline to the last visit. In order to perform the statistical analysis, the Snellen Visual Acuity was converted into an approximate ETDRS letter score. Two-side, p-values < 0.05 were considered statistically significant.

3. Results

3.1. General

More than 1500 SD-OCT images were analyzed by one of the authors (D.I.), followed by a more detailed analysis for 210 SD-OCT images performed by the same author.

A total of 18 eyes belonging to 13 patients (4 (30.8%) females, 9 (69.2%) males) retrospectively diagnosed with AVLs were included in the study. The mean age at presentation was: 79.9 ± 7.1 years (range 67–91). Six patients (33.3%) had bilateral AVLs. The mean ETDRS score at baseline was 69 ± 18 letters.

At baseline, 16/18 (88.9%) eyes presented a vitelliform stage while 2/18 (11.1%) presented a vitelliruptive stage. In terms of the OCT features found in our study eyes containing AVLs, 9 eyes (50%) had associated SDD, 8 eyes (44.4.%) had cuticular drusen, 4 eyes (22.2%) had soft drusen, 9 eyes (50%) had HRF, 5 eyes (27.8%) had SRF, 2 eyes (11.1%) had PED, and 3 eyes (16.6%) had no other findings, except the vitelliform lesion. The qualitative analysis showed a continuous ELM in 10/18 eyes (55.5%), a discontinuous ELM in 6/18 eyes (33.3%) and an indistinguishable ELM in 2/18 eyes (11.1%). EZ was continuous in 4/18 eyes (22.2%), discontinuous in 13/18 eyes (72.2%) and indistinguishable in 1/18 eyes (5.5%) (see Table 1 and Figure 4).

Table 1. OCT features of patients with AVLs at baseline and last visit.

Stage of the Disease, n (%)	Baseline	Last Visit
Vitelliform	16/18 (88.9)	7/9 (77.8)
Pseudohypopyon	0/18 (0)	0/9 (0)
Vitelliruptive	2/18 (11.1)	2/9 (22.2)
Atrophic	0/18 (0)	0/9 (0)
SDD	9/18 (50)	4/9 (44.4)

Table 1. *Cont.*

Stage of the Disease, n (%)	Baseline	Last Visit
Soft drusen	4/18 (22.2)	8/9 (88.8)
Cuticular drusen	8/18 (44.4)	4/9 (44.4)
SRF associated with AVL	5/18 (27.8)	0/9 (0)
HRF	9/18 (50)	3/9 (33.3)
PED	2/18 (11.1)	5/9 (55.5)
ELM		
Continuous	10/18 (55.5)	3/9 (33.3)
Discontinuous	6/18 (33.3)	2/9 (22.2)
Indistinguishable	2/18 (11.1)	4/9 (44.4)
EZ		
Continuous	4/18 (22.2)	0/9 (0)
Discontinuous	13/18 (72.2)	7/9 (77.8)
Indistinguishable	1/18 (5.5)	2/9 (22.2)

n = number; SDD = subretinal drusenoid deposit, SRF = subretinal fluid; AVL = acquired vitelliform lesion; HRF = hyperreflective foci; PED = pigment epithelium detachment; ELM = external limiting membrane; EZ = ellipsoid zone.

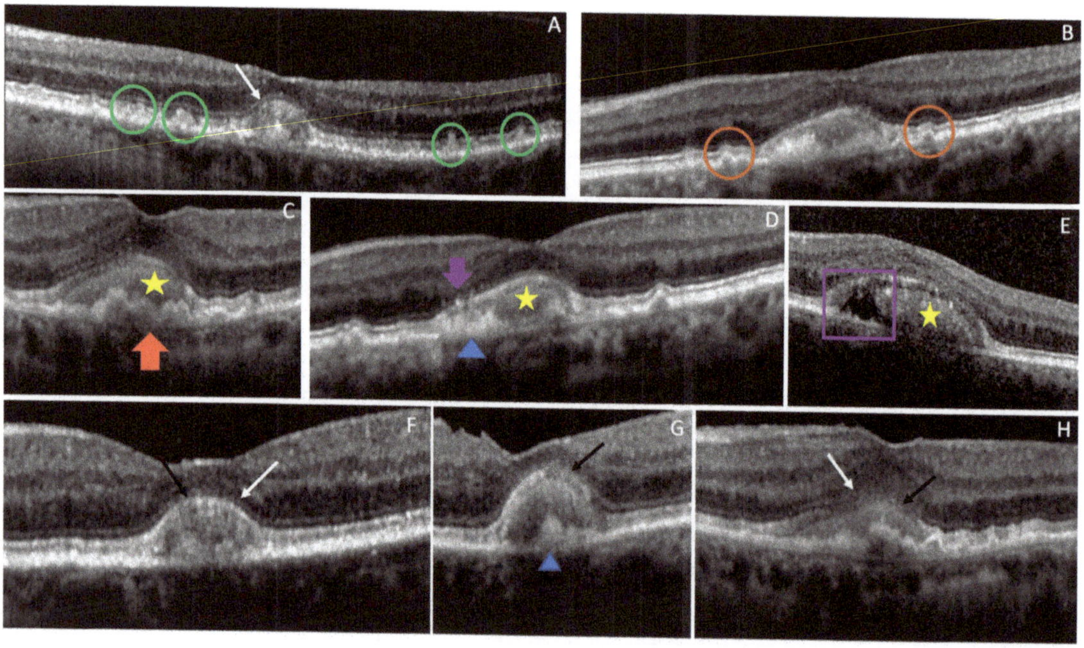

Figure 4. Associated retinal lesions. SD-OCT B-scans passing through the AVLs of clinical study eyes; (**A**) the green circles show SDD and the white arrow continuous ELM; (**B**) the brown circles show cuticular drusen; (**C**) the red arrow shows a drusenoid PED and the yellow star the vitelliform deposit; (**D**) the yellow star shows the vitelliform deposit, the purple arrow an HRF, the blue arrow head the RPE thickening; (**E**) the yellow star shows the vitelliform deposit, the purple square the SRF; (**F**) the white arrow indicates a continuous ELM and the black arrow a continuous EZ; (**G**): the black arrow indicates a discontinuous EZ and the blue arrow head the RPE thickening; (**H**) the white arrow indicates a indistinguishable ELM and the black arrow an indistinguishable EZ.

Four patients/7 eyes (43.75%) received no follow-up. A total of 7 patients/9 eyes (56.25%) received an ophthalmological follow-up. The mean follow-up duration was: 23.33 ± 20.63 months (range 5–56 months). Six patients/six eyes with a well-defined AVL were further included in a specific volumetric analysis.

Seven eyes were treated with anti-VEGF with a mean of 3 injections/eye (min 1–max 8).

3.2. Vitelliform Lesions Analysis

At baseline, the average maximum height was 225.73 ± 336.33 μm (range 37.09–1211.60). The average base width size was 622.40 ± 419.71 μm (range 103.31–1855.37). The measured average perimeter was 1648.67 ± 943.90 μm (range 706.05–4274.17) while the average area was 17,803.28 ± 10,055.30 μm^2 (range 7012.07–45,624.89) (See Table 2). The density of the lesions was further analyzed. The average mean gray value was 140.49 ± 16.54 (range 101.71–170.53), the average modal gray value was 130.37 ± 25.62 (range 78.86–165.22), the average min gray value was 65.30 ± 20.73 (range 20.29–104.24), the average maximum gray value was 212.74 ± 20.61 (range 175.43–248.78), the average integrated density was 2,366,898.95 ± 1,071,937 (range 892,257.21–4,640,455.74) and the average RawIntden was 20,614.39 ± 9630.70 (range 9034.45–40,785.26) (see Table 2).

Table 2. Vitelliform lesion characteristics at baseline.

	Average ± SD	Median	Min	Max
Vitelliform Lesion Size				
Maximum height (μm)	225.73 ± 336.33	114.69	37.09	1211.60
Base width (μm)	622.40 ± 419.71	556.00	103.31	1855.37
Area (μm^2)	17,803.28 ± 10,055.30	14,579.10	7012.07	45,624.89
Perimeter (μm)	1648.67 ± 943.90	1277.64	706.05	4274.17
Vitelliform lesion density				
Mean gray value	140.49 ± 16.54	143.39	101.71	170.53
Modal gray value	130.37 ± 25.62	132.54	78.86	165.22
Min gray level	65.30 ± 20.73	63.79	20.29	104.24
Max gray level	212.74 ± 20.61	211.64	175.43	248.78
Integrated density	2,366,898.95 ± 1,071,937.10	2,072,612.71	892,257.21	4,640,455.74
RawIntden	20,614.39 ± 9630.70	17,422.93	9034.45	40,785.26

SD = standard deviation; Min = minimum; Max = maximum; μm = micrometer. RawIntden = Raw integrated density.

3.3. Retinal Layer Thickness

The different retinal layer thickness was assessed in the central 1mm subfield for the vitelliform and control groups (see supplementary Tables S1 and S2).

We compared the average values for the measured retinal layer thickness in the vitelliform and control groups. The RPE thickness in the central 1 mm (45.89 ± 27.84 μm vs. 15.57 ± 1.40 μm) as well as the RPE central max (101.67 ± 53.91 μm vs. 23.363.63 μm) were significantly increased in the vitelliform group compared to the control group ($p < 0.001$ and $p < 0.001$). The thickness in the central 1 mm (122.72 ± 36.09 μm vs. 86.43 ± 4.57 μm) and ORL central max (179.44 ± 56.83 μm vs. 97.36 ± 6.66 μm) were also significantly increased in the vitelliform group compared to the control ($p < 0.001$ and $p < 0.001$). ONL 1 mm was significantly decreased in the vitelliform group compared to the control: 77.94 ± 18.30 μm vs. 88.64 ± 7.65 μm ($p = 0.049$). The CRT in the central 1mm subfield (301.94 ± 50.02 μm vs. 258.43 ± 17.16 μm), the CRT min (252.94 ± 53.80 μm vs. 216.93 ± 19.87 μm) as well as the CRT max (345.19 ± 47.63 μm vs. 307.57 ± 16.10 μm) were significantly increased in the vitelliform group compared to the control ($p = 0.005$, $p = 0.025$ and $p = 0.009$) (see Table 3).

Table 3. Comparing retinal layer thickness between vitelliform and control groups.

Average Retinal Layer Thickness (μm)	Vitelliform Group ± SD	Control Group ± SD	*p*
RPE 1 mm	45.89 ± 27.84	15.57 ± 1.40	<0.001
RPE central min	10.17 ± 5.37	10.71 ± 2.23	0.723
RPE central max	101.67 ± 53.91	23.36 ± 3.63	<0.001
ONL 1 mm	77.94 ± 18.30	88.64 ± 7.65	0.049
ONL central min	45.33 ± 20.34	53.86 ± 16.11	0.208
ONL central max	115.72 ± 21.84	112.14 ± 9.69	0.573
ORL 1 mm	122.72 ± 36.09	86.43 ± 4.57	<0.001
ORL central min	80.00 ± 11.08	78.14 ± 7.39	0.401 *
ORL central max	179.44 ± 56.83	97.36 ± 6.66	<0.001
IRL 1 mm	180.83 ± 30.89	172.14 ± 15.24	0.352
IRL central min	101.67 ± 40.35	121.36 ± 15.90	0.174
IRL central max	241.11 ± 35.81	216.29 ± 36.28	0.105
Retina 1 mm	303.00 ± 50.54	258.43 ± 17.16	0.004
Retina central min	254.17 ± 55.49	216.93 ± 19.87	0.023
Retina central max	348.17 ± 48.23	307.57 ± 16.10	0.005

* = Mann–Whitney U test. μm = micrometer; mm = millimeter; RPE = retinal pigment epithelium; min = minimum; max = maximum; ONL = outer nuclear layer; ORL = outer retinal layer; IRL = inner retinal layer, CRT = central retinal thickness. *p* indicates statistical significance.

3.4. Vitelliform Lesions in Evolution

For six patients, a secondary analysis was carried out in order to investigate the evolution of the vitelliform lesions' volume in time (see Table 4). The mean number of SD-OCT macular scans per eye used for the vitelliform volume plots was 9.22 ± 6.12 (range 3–25). In two eyes the volume increased while in four eyes it decreased. The difference between the mean vitelliform lesion's volume at baseline and its volume at the last visit was not statistically significant ($p = 0.725$).

Table 4. Vitelliform lesion volume.

	Vitelliform Lesion Volume (μm³)		
Eye	Volume Baseline (μm³)	Volume Last Visit (μm³)	Difference 2–1 (μm³)
1	30,064,897.20	55,831,556.40	25,766,659.20
2	32,579,991.60	15,413,652.00	−17,166,339.60
3	4,310,901.60	20,337,099.60	16,026,198.00
4	13,922,006.40	9,493,209.60	−4,428,796.80
5	20,519,072.40	18,330,301.20	−2,188,771.20
6	17,674,183.20	17,100,080.40	−574,102.80
Average	19,845,175.40	22,750,983.20	2,905,807.80
Median	19,096,627.80	18,330,301.20	−574,102.80
Min	4,310,901.60	9,493,209.60	−17,166,339.60
Max	32,579,991.60	55,831,556.40	25,766,659.20

Further on, we calculated the difference between the baseline and the last visit in terms of vitelliform lesion size (see Table 5), density (see Supplementary Table S3), as well as RPE, ONL, ORL, IRL and CRT (see Table 6). The maximum height decreased in one eye while in five it increased ($p = 0.232$). The base width increased in three and decreased in three eyes ($p = 0.125$). The area and perimeter decreased in two eyes while in four it increased ($p = 0.103$ and $p = 0.081$).

Table 5. Difference between baseline and last visit vitelliform lesion size.

	Maximum Height (µm)	Base Width (µm)	Area (µm^3)	Perimeter (µm)
Average ± SD	−95.46 ± 296.04	248.27 ± 467.70	3398.72 ± 5733.13	327.48 ± 490.86
Median	9.51	21.68	1313.74	154.15
Min	−696.30	−83.26	−2509.03	−191.55
Max	87.32	1096.28	11,547.57	990.35

The mean density of the lesion decreased in 3 eyes ($p = 0.350$), while the max gray level ($p = 0.236$) and the min gray level decreased in 4 eyes ($p = 0.286$). Eyes that had a decreased mean gray value presented decreased integrated density ($p = 0.269$) and RawIntdensity values ($p = 0.232$) (see Supplementary Table S3).

Differences between the first and last presentation in the vitelliform group in terms of the retinal layer thickness are displayed in Table 6. The average RPE thickness, CRT and ORL for the 1 mm, min and max decreased at the last visit, but the differences were not significant ($p > 0.05$) (see Table 6).

Table 6. Difference in retinal layer thickness between baseline and last visit in vitelliform group.

	Average ± SD	p
RPE 1 mm	−3.11 ± 22.87	0.347
RPE min	−2.67 ± 3.77	0.067
RPE max	−5.67 ± 61.03	0.393
ONL 1 mm	1.22 ± 15.67	0.821
ONL min	−5.56 ± 11.70	0.192
ONL max	22.56 ± 52.62	0.234
CRT 1mm	−14.44 ± 41.76	0.329
CRT min	−8.00 ± 48.73	0.635
CRT max	−18.33 ± 35.70	0.362
ORL 1 mm	−7.11 ± 60.87	0.379
ORL min	−6.22 ± 10.51	0.113
ORL max	−7.11 ± 60.87	0.735
IRL 1 mm	1.67 ± 15.48	0.754
IRL min	1.00 ± 30.19	0.923
IRL max	13.67 ± 43.99	0.378

µm = micrometer; mm = millimeter; RPE = retinal pigment epithelium; min = minimum; max = maximum; ONL = outer nuclear layer; ORL = outer retinal layer; IRL = inner retinal layer, CRT = central retinal thickness.

At the last visit, 7/9 eyes (77.8%) presented with the vitelliform stage, while 2 of them (22.2%) presented with the vitelliruptive stage. The status of ELM was as follows: 3/9 (33.3%) continuous, 2/9 (22.2%) discontinuous and 4/9 eyes (44.4%) indistinguishable. EZ was discontinuous in 7/9 eyes (77.8%) and indistinguishable in 2/9 eyes (22.2%). In terms of the AVL's fate, the vitelliform lesion increased in 7/9 eyes (77.8%), was stable with minimal change in 1/9 (11.1%) and in 1/9 eyes (11.1%) it had resorbed with a superior collapsed area.

We further illustrated 2 representative cases. An 85-year-old patient, with a vitelliform lesion in the RE, with a BCVA of 80 ETDRS letters at baseline (see Figure 5A).

After 25 months and 4 anti-VEGF Bevacizumab injections, the lesion increased in size (see Figure 5B). In the next 14 months, he received another 2 anti-VEGF injections with Bevacizumab and 2 anti-VEGF injections with Aflibercept (see Figure 5C). The lesion transformed into vitelliruptive while the BCVA was 70 ETDRS letters (see Figure 5E).

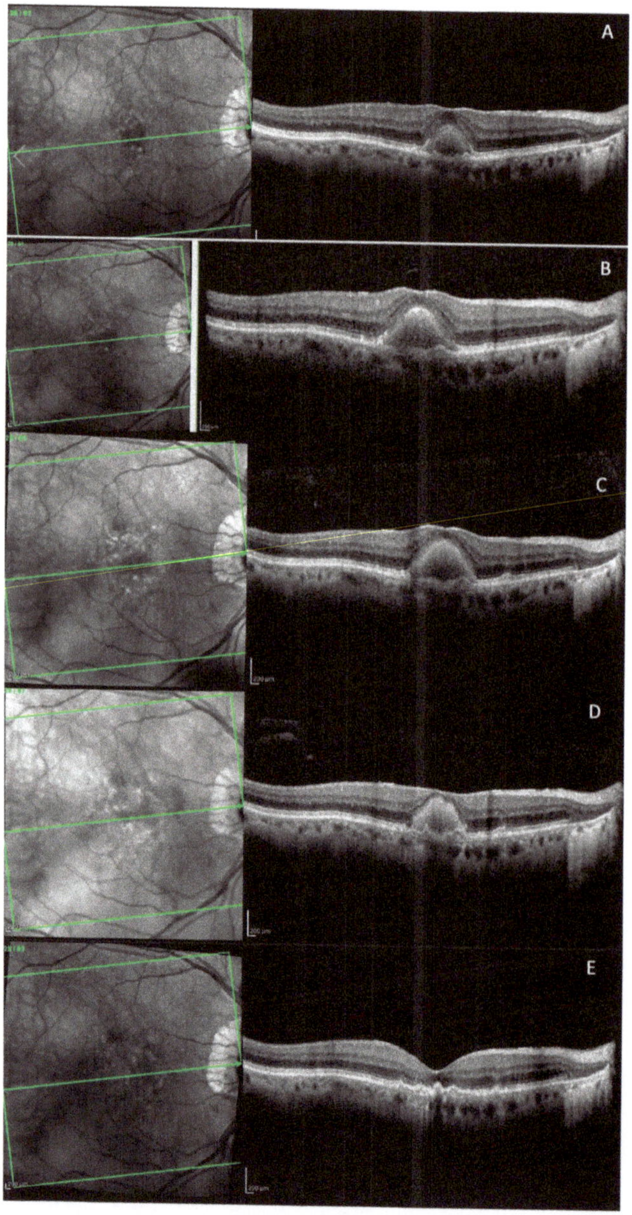

Figure 5. Spectral-domain OCT (SD-OCT) in a patient with AVL presenting the vitelliform stage at baseline (**A**); after 4 anti-VEGF injections with Bevacizumab, respectively 25 months later, the lesion increased (**B**); after another 2 anti-VEGF injections with Bevacizumab and 2 injections with Aflibercept (**C**), respectively, after another 14 months; after another 2 months the lesion started to shrink (**D**); after another month the lesion resorbed (**E**).

Figure 6 presents the case of a 71-year-old patient with a small vitelliform lesion in his LE, with a BCVA ETDRS of 75 letters (A). Forty-three months later, the lesion increased, with a BCVA ETDRS of 75 letters (B).

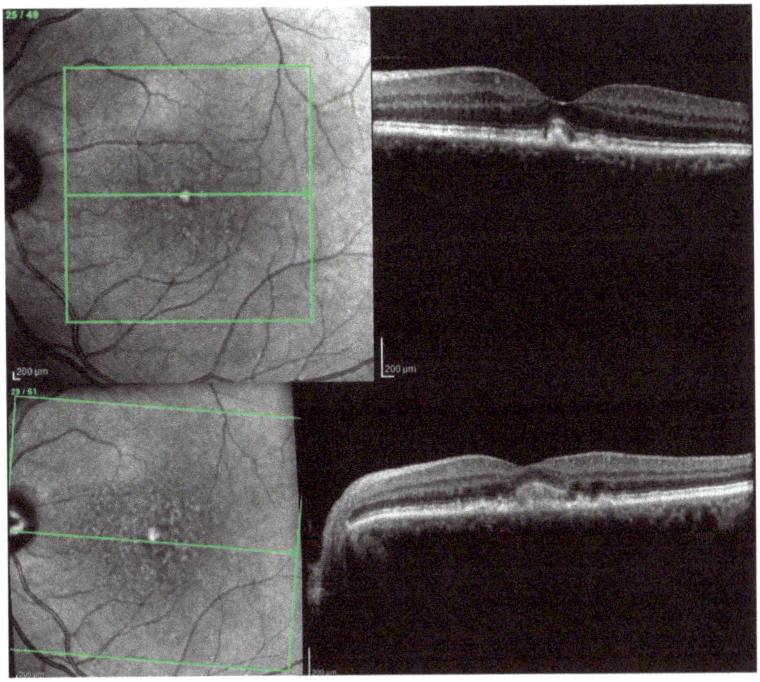

Figure 6. Spectral-domain OCT (SD-OCT) in patient with AVL presenting the vitelliform stage at baseline (**upper image**) and 43 months later at last visit (**lower image**).

Seven eyes (43.75%) were treated with intravitreal anti-VEGF agent injections. The initial and final BCVA is also provided along with the mean number of injections: 3/eye (min 1–max 8), and type of anti-VEGF agent: 16 Bevacizumab and 4 Aflibercept. Overall, the BCVA worsened between the baseline and final visits by −6.43 ± 9 letters, but the difference did not reach statistical significance ($p = 0.054$) (see Table 7).

Table 7. Eyes with vitelliform lesions treated with anti-VEGF injections.

No.	Initial BCVA (Letters)	Follow-Up (Months)	Anti-VEGF Injections (No.)	Final BCVA (Letters)	Difference BCVA (Letters)
1.	80	29	8 (6B + 2A)	70	−10
2.	80	29	5 B	55	−25
3.	70	6	2 B	65	−5
4.	80	56	1A	80	0
5.	75	56	1 B	75	0
6.	85	5	2 B	80	−5
7.	25	11	1 A	25	0
average ± SD	70 ± 20.7	27.4 ± 21.87	2.9 ± 2.67	64.3 ± 19.46	−6.43 ± 9

BCVA = Best corrected visual acuity; No. = number; anti-VEGF = anti-vascular endothelial growth-factor; B = Bevacizumab; A = Aflibercept.

4. Discussion

In the present study, we used SD-OCT to characterize the anatomy of AVLs in terms of their dimension, density and relation with retinal layers.

In our cohort, the AVLs were more frequent in patients in their seventh–eighth decade (79.25 ± 6.7 years). The age was similar to that reported in other studies 78.6 ± 6 years [18], as well as 72 years by Juliano J. et al. [7]. A slight male predominance was observed in our cohort 69.2% vs. 30.8% ($p = 0.019$), similar to the previously reported one: 56.25% vs. 43.75% [7].

In our study group, 33.3% had bilateral AVLs, similar to 31.25% found by Juliano J. et al. [7]. Saade C. et al. performed a review regarding AVLs in AMD patients and found 19% bilateral AVLs in a retrospective study with 19 eyes belonging to 16 patients, 44% bilateral AVLs in 13 eyes of 9 patients and 33% bilateral AVLs in 32 eyes of 24 patients [19].

We confirmed the presence of cuticular drusen in 10 eyes (62.5%), while others noted a prevalence of AVLs in patients with cuticular drusen from 1.2% to 24.2% [20]. The same author observed that the association between the AVL and cuticular drusen seems to lead to a collapse with the resolution of the vitelliform material, doubled by RPE failure and atrophy with the disappearance of drusen [20]. Contrary to another study in which all the 32 eyes belonging to 24 patients presented serous PED [21], we found only 2 eyes (12.5%) with RPE detachment. Hyperreflective foci (HRF) were present in 5 eyes (31.25%) in our series, while other authors found HRF in 9.8% of cases, of which five eyes also presented SRF (25%) [6].

Previous authors observed that ELM remains intact in most cases until complete AVL collapse and the subsequent development of the RPE and outer retinal atrophy [5]. Our study showed a continuous ELM in 55.5% compared to a continuous EZ in 22.2% of eyes. Based on our findings, it seems that the majority of eyes presented at baseline a discontinuous EZ and a continuous ELM. The discontinuity of the EZ could be explained by the findings from adaptive optics scanning laser ophthalmoscopy studies, which suggested that a rarefaction of cones doubled by morphological changes were found within the vitelliform lesion [22].

When analyzing the vitelliform lesion size, it seems that the base width was constantly larger compared to the maximum height. For both parameters, the range was large, which underlines the great variability between lesions: the average maximum height was 225.73 ± 335.33 µm (range 37.09–1211.60) and the average base width size was 622.40 ± 419.71 µm (range 103.31–1855.37). It was previously shown in an adjusted analysis that the baseline factors associated with an increased risk of MA included the base width (HR, 1.22; 95% CI, 1.16–1.28) and maximum height (HR, 2.61; 95% CI, 1.82–3.74) [23]. The perimeter and area analysis could further help in better describing the vitelliform lesions at the baseline and over time.

Similar values for the average and modal gray demonstrated a strong similarity between the density of the vitelliform lesions in our study group. One explanation could be the same disease stage, more precisely the vitelliform one. The average mean gray value was 140.49 ± 16.54 (range 101.71–170.53) and the average modal gray value was 130.37 ± 25.62 (range 78.86–165.22). It is noteworthy that to the best of our knowledge, this study is the first tentative study of density measurements for vitelliform lesions. This information helps to better describe changes within the lesion as it progresses. Moreover, it could represent a possible biomarker of disease progression. We also used a specific density marker called integrated density, which is related to the area in which the density was measured. On the same note, Boiko et al. investigated the retinal tissue area (RTA) and optical density (ODRT) of the retinal optical slice portion located in the central subfield, as well as their ratio (RTA/ODRT) in the presence of diabetic macular edema (DME) or of intraretinal cystic fluid (IRF) in neovascular age-related macular degeneration (nAMD) before and after anti-VEGF treatment [24]. The authors found that all the above-mentioned markers could be used as predictors of functional and anatomical outcomes since the RTA correlated positively both with post-anti-VEGF retinal thickness ($r = 0.76$; $p < 0.001$) and BCVA ($r = 0.67$; $p < 0.001$), while ODRT was moderately negatively correlated ($r = 20.26$;

$p = 0.049$ and r = 20.48; $p = 0.001$, respectively) [24]. RTA/ODRT was strongly positively correlated (r = 0.75; $p < 0.001$ and r = 0.85; $p < 0.001$, respectively) [24].

The RPE, ORL thickness and CRT in the vitelliform group at baseline were increased compared to the control. This is contrary to other authors who found a similar RPE thickness with the control group, but they compared it with BVMD eyes [25]. As previously shown by other authors, the hyperreflective focal thickenings of the RPE band seen in SD-OCT may represent RPE hyperplasia or melanolipofuscin granules in the macrophages [14]. In terms of the RPE evolution, first it undergoes hypertrophy and then disruption, which could explain the increased RPE thickness values found in our study group. The ORL and CRT increase could mirror the RPE's increased thickness. Agreeing with other descriptions, the central ONL thickness and min ONL thickness were decreased in our study, which could show the impact of the vitelliform lesion upon the PR and further on over the BCVA. As observed by another group, the ONL thickness was significantly lower compared to the control group and it correlated with the BCVA [26]. Other authors identified a photoreceptor equivalent thickness of 6.5 μm thicker in the BVMD compared to control [25]. The IRL 1mm and central max values were increased compared to the control group but were not statistically significant. Based on our findings, it seems that together with the RPE, the ORL layer and CRT are characterized by increased values. On the contrary, the ONL showed decreased values, as a proof of the vitelliform lesion impact on the PR.

All parameters analyzed in evolution have demonstrated increases as well as decreases in values, without a definite trend. We found a similar pattern for the density, retinal layer thickness, vertical height, horizontal diameter, area or perimeter, yet the changes were not significant. Previous authors suggested that during the disease progression, a change in the lesion's composition or density might be possible, which further on could be responsible for the change in its overall reflectivity, from hyperreflective to hyporeflective [14]. Other authors found a significant shrinkage from 142.5 to 125.5 ($p = 0.009$) when comparing the initial versus final lesion height values in the AVL group [4]. However, from our data it was not possible to ascertain a significant decrease.

Data found by another study suggest that there might be an increase in the lesion's size from the vitelliform stage to pseudohypopyon, while from the pseudohypopyon to the vitelliruptive stage there is a decrease in the lesion dimension, which could be a plausible explanation for our results [14]. In the same study, the authors found that during the disease evolution through different stages, when the lesion decreases, there is also a resorption of the vitelliform material doubled by a change in the lesion reflectivity: from hyperreflective to either mixed hyper + hyporeflective or hyporeflective, and also a photoreceptor loss [14]. The same authors raised the question about what triggers the fragmentation and resorption of the vitelliform material, which seems to remain unanswered [14].

At the final visit, the status of the ELM was discontinuous in 2/9 (22.2%) and indistinguishable in 4/9 eyes (44.4%), while the EZ was discontinuous in 7/9 eyes (77.8%) and indistinguishable in 2/9 eyes (22.2%).

Regarding the AVL fate, most of the eyes demonstrated an increase in the vitelliform lesion (77.8%), 11.1% showed a stable lesion with minimal change and 11.1% progressed to resorption with a superior collapsed area.

When we analyzed the BCVA's evolution in the eyes that received anti-VEGF agents, we found a negative trend but this was not statistically significant: -6.43 ± 9 letters ($p = 0.054$). Based on our findings, it seems that there is no benefit of anti-VEGF injections in eyes with vitelliform lesions, but larger studies need be performed in order to confirm this finding. These data are in accordance with those coming from other studies, which showed a significant mean BCVA reduction from the baseline to the last visit of at least 2 lines [2]. Another study underlined a visual acuity better or equal to 0.6 at presentation in 43% of the eyes that progressively decreased to 8% over 10 years of follow-up [27]. When the size of the lesion was analyzed in conjunction with the BCVA, a low correlation was found, so the authors speculated that BCVA seems to be less affected by quantitative changes (such as lesions size) and more dependent on qualitative changes such as the

IS/OS interface integrity or reflectivity changes, since they found progression in the central photoreceptor IS/OS interface status and changes in the lesion reflectivity in the eyes with a significant mean BCVA reduction from baseline to the last visit [14]. These findings are in accordance with data from another study who found that the BCVA at baseline was best predicted by the IS/OS junction (p = 0.0002) and subfoveal integrity of the ELM (p = 0.001) in SD-OCT [28].

We acknowledge several limitations of our study. Firstly, the study is retrospective. Secondly, the sample size was small, limiting our ability to detect other statistical correlations. Thirdly, the genetic data to better differentiate between AOVMD, AVL and AMD are missing. Fourthly, the frequency of the follow-up between patients was variable. A complete automated method for vitelliform lesion identification and measurement would be useful in the future to decrease any potential biases from manual delineation.

5. Conclusions

In our small series of AVLs in adult patients, we found increased RPE thickness as a sign of hyperplasia, contrary to the central and min ONL layer which were decreased, mirroring the impact of the vitelliform lesion over the PR. Most of the analyzed eyes presented a discontinuous EZ, both at baseline and in evolution, being a sign of visual impairment. During the follow-up some of the AVLs had increased while others regressed, which was translated in divergent results regarding height, width, area, perimeter or volume between the eyes. The density analysis helps us better evaluate the AVLs' evolution in time, since vitelliform resorption is further translated into a reflectivity or density change.

Eyes that received anti-VEGF injections did not show signs of improvement regarding the BCVA, since the mean change in our study group was -6.43 ± 9 letters (p = 0.054).

Supplementary Materials: The following supporting information can be downloaded at: https://www.mdpi.com/article/10.3390/biomedicines11051382/s1, Table S1: Retinal layer thickness in vitelliform group; Table S2: Retinal layer thickness in control group; Table S3: Difference between baseline and last visit vitelliform lesion density.

Author Contributions: Conceptualization, I.D., S.-D.N., G.-A.M. and L.-B.G.-H.; methodology, I.D., S.-D.N., G.-A.M. and L.-B.G.-H.; software, I.D.; validation, I.D., S.-D.N., G.-A.M. and L.-B.G.-H.; formal analysis, I.D.; investigation, I.D.; resources, I.D.; data curation, I.D., G.-A.M.; S.-D.N., L.-B.G.-H.; writing—original draft preparation, I.D.; writing—review and editing, I.D.; S.-D.N., G.-A.M., L.-B.G.-H.; visualization, I.D.; supervision, S.-D.N.; project administration, I.D.; funding acquisition, I.D. All authors have read and agreed to the published version of the manuscript.

Funding: This work was supported by a grant of the Ministry of Research, Innovation and Digitization, CCCDI-UEFISCDI, project number PN-III-P2-2.1-PED-2021-2709, within PNCDI III.

Institutional Review Board Statement: The study was conducted in accordance with the Declaration of Helsinki and approved by the Ethics Committee) of "IULIU HATIEGANU" UNIVERSITY OF MEDICINE AND PHARMACY (IHUMP), Cluj-Napoca, Romania (no.273/19.09.2022).

Informed Consent Statement: Patient consent was waived due to the retrospective and observational nature of the study.

Data Availability Statement: The data presented in this study are available upon request from the corresponding author. The data are not publicly available due to privacy reasons.

Conflicts of Interest: The authors declare no conflict of interest.

References

1. Arnold, J.J.; Sarks, J.; Killingsworth, M.C.; Kettle, E.; Sarks, S.H. Adult vitelliform macular degeneration: A clinicopathological study. *Eye* **2003**, *17*, 717–726. [CrossRef] [PubMed]
2. Greaves, A.H.; Sarks, J.P.; Sarks, S.H. Adult vitelliform macular degeneration: A clinical spectrum. *Aust. N. Z. J. Ophthalmol.* **1990**, *18*, 171–178. [CrossRef] [PubMed]

3. Balaratnasingam, C.; Hoang, Q.V.; Inoue, M.; Curcio, C.A.; Dolz-Marco, R.; Yannuzzi, N.A.; Dhrami-Gavazi, E.; Yannuzzi, L.A.; Freund, K.B. Clinical Characteristics, Choroidal Neovascularization, and Predictors of Visual Outcomes in Acquired Vitelliform Lesions. *Am. J. Ophthalmol.* **2016**, *172*, 28–38. [CrossRef] [PubMed]
4. Rocha Bastos, R.; Ferreira, C.S.; Brandão, E.; Falcão-Reis, F.; Carneiro Â, M. Multimodal Image Analysis in Acquired Vitelliform Lesions and Adult-Onset Foveomacular Vitelliform Dystrophy. *J. Ophthalmol.* **2016**, *2016*, 6037537. [CrossRef] [PubMed]
5. Brinkmann, M.; Bacci, T.; Kar, D.; Messinger, J.D.; Sloan, K.R.; Chen, L.; Hamann, T.; Wiest, M.; Freund, K.B.; Zweifel, S.; et al. Histology and Clinical Lifecycle of Acquired Vitelliform Lesion, a Pathway to Advanced Age-Related Macular Degeneration. *Am. J. Ophthalmol.* **2022**, *240*, 99–114. [CrossRef]
6. Chen, K.C.; Jung, J.J.; Curcio, C.A.; Balaratnasingam, C.; Gallego-Pinazo, R.; Dolz-Marco, R.; Freund, K.B.; Yannuzzi, L.A. Intraretinal Hyperreflective Foci in Acquired Vitelliform Lesions of the Macula: Clinical and Histologic Study. *Am. J. Ophthalmol.* **2016**, *164*, 89–98. [CrossRef]
7. Juliano, J.; Patel, S.; Ameri, H. Acquired Vitelliform Macular Degeneration: Characteristics and Challenges of Managing Subretinal Fluid. *J. Ophthalmic Vis. Res.* **2021**, *16*, 582–591. [CrossRef]
8. Krämer, F.; White, K.; Pauleikhoff, D.; Gehrig, A.; Passmore, L.; Rivera, A.; Rudolph, G.; Kellner, U.; Andrassi, M.; Lorenz, B.; et al. Mutations in the VMD2 gene are associated with juvenile-onset vitelliform macular dystrophy (Best disease) and adult vitelliform macular dystrophy but not age-related macular degeneration. *Eur. J. Hum. Genet.* **2000**, *8*, 286–292. [CrossRef]
9. Joshi, K.M.; Nesper, P.L.; Fawzi, A.A.; Mirza, R.G. Optical coherence tomography angiography in adult-onset foveomacular vitelliform dystrophy. *Retina* **2018**, *38*, 600–605. [CrossRef]
10. Crincoli, E.; Zhao, Z.; Querques, G.; Sacconi, R.; Carlà, M.M.; Giannuzzi, F.; Ferrara, S.; Ribarich, N.; L'abbate, G.; Rizzo, S.; et al. Deep learning to distinguish Best vitelliform macular dystrophy (BVMD) from adult-onset vitelliform macular degeneration (AVMD). *Sci. Rep.* **2022**, *12*, 12745. [CrossRef]
11. EyeEiki. Available online: https://eyewiki.aao.org/Best_Disease_and_Bestrophinopathies (accessed on 20 September 2022).
12. Lotery, A.J.; Munier, F.L.; Fishman, G.A.; Weleber, R.G.; Jacobson, S.G.; Affatigato, L.M.; Nichols, B.E.; Schorderet, D.F.; Sheffield, V.C.; Stone, E.M. Allelic variation in the VMD2 gene in best disease and age-related macular degeneration. *Investig. Ophthalmol. Vis. Sci.* **2000**, *41*, 1291–1296.
13. Tripathy, K.; Salini, B. *Best Disease*; StatPearls Publishing: St. Petersburg, FL, USA, 2023.
14. Querques, G.; Forte, R.; Querques, L.; Massamba, N.; Souied, E.H. Natural course of adult-onset foveomacular vitelliform dystrophy: A spectral-domain optical coherence tomography analysis. *Am. J. Ophthalmol.* **2011**, *152*, 304–313. [CrossRef] [PubMed]
15. Midena, E.; Torresin, T.; Velotta, E.; Pilotto, E.; Parrozzani, R.; Frizziero, L. OCT Hyperreflective Retinal Foci in Diabetic Retinopathy: A Semi-Automatic Detection Comparative Study. *Front. Immunol.* **2021**, *12*, 613051. [CrossRef] [PubMed]
16. Schindelin, J.; Arganda-Carreras, I.; Frise, E.; Kaynig, V.; Longair, M.; Pietzsch, T.; Preibisch, S.; Rueden, C.; Saalfeld, S.; Schmid, B.; et al. Fiji: An open-source platform for biological-image analysis. *Nat. Methods* **2012**, *9*, 676–682. [CrossRef] [PubMed]
17. Sigal, I.A.; Schuman, J.S.; Ishikawa, H.; Kagemann, L.; Wollstein, G. A Problem of Proportions in OCT-Based Morphometry and a Proposed Solution. *Investig. Opthalmology Vis. Sci.* **2016**, *57*, 484–485. [CrossRef]
18. Wilde, C.; Awad, M.; Giannouladis, K.; Lakshmanan, A.; Yeung, A.M.-H.; Dua, H.; Amoaku, W.M.K. Natural course of adult-onset vitelliform lesions in eyes with and without comorbid subretinal drusenoid deposits. *Int. Ophthalmol.* **2020**, *40*, 1501–1508. [CrossRef]
19. Saade, C.; Greenberg, P.B. Acquired Vitelliform Lesions in Age-Related Macular Degeneration. *Investig. Ophthalmol. Vis. Sci.* **2016**, *57*, 2061.
20. Fragiotta, S.; Fernández-Avellaneda, P.; Breazzano, M.P.; Scuderi, G. Clinical Manifestations of Cuticular Drusen: Current Perspectives. *Clin. Ophthalmol.* **2021**, *15*, 3877–3887. [CrossRef]
21. Saito, M.; Iida, T.; Freund, K.B.; Kano, M.; Yannuzzi, L.A. Clinical Findings of Acquired Vitelliform Lesions Associated with Retinal Pigment Epithelial Detachments. *Am. J. Ophthalmol.* **2014**, *157*, 355–365.e2. [CrossRef]
22. Scoles, D.; Sulai, Y.N.; Cooper, R.F.; Higgins, B.P.; Johnson, R.D.; Carroll, J.; Dubra, A.; Stepien, K.E. Photoreceptor inner segment morphology in best vitelliform macular dystrophy. *Retina* **2017**, *37*, 741–748. [CrossRef]
23. Chandra, S.; Gurudas, S.; Narayan, A.; Sivaprasad, S. Incidence and Risk Factors for Macular Atrophy in Acquired Vitelliform Lesions. *Ophthalmol. Retin.* **2021**, *6*, 196–204. [CrossRef] [PubMed]
24. Boiko, E.V.; Maltsev, D.S. Quantitative optical coherence tomography analysis of retinal degenerative changes in diabetic macular edema and neovascular age-related macular degeneration. *Retina* **2018**, *38*, 1324–1330. [CrossRef]
25. Kay, C.N.; Abràmoff, M.D.; Mullins, R.F.; Kinnick, T.R.; Lee, K.; Eyestone, M.E.; Chung, M.M.; Sohn, E.H.; Stone, E.M. Three-dimensional Distribution of the Vitelliform Lesion, Photoreceptors, and Retinal Pigment Epithelium in the Macula of Patients With Best Vitelliform Macular Dystrophy. *JAMA Ophthalmol.* **2012**, *130*, 357–364. [CrossRef] [PubMed]
26. Augstburger, E.; Orès, R.; Mohand-Said, S.; Mrejen, S.; Keilani, C.; Antonio, A.; Condroyer, C.; Andrieu, C.; Sahel, J.-A.; Zeitz, C.; et al. Outer Retinal Alterations Associated with Visual Outcomes in Best Vitelliform Macular Dystrophy. *Am. J. Ophthalmol.* **2019**, *208*, 429–437. [CrossRef] [PubMed]

27. Glacet-Bernard, A.; Soubrane, G.; Coscas, G. Dégénérescence maculaire vitelliforme de l'adulte. Etude rétrospective d'une série de 85 patients [Macular vitelliform degeneration in adults. Retrospective study of a series of 85 patients]. *J. Fr. Ophtalmol.* **1990**, *13*, 407–420.
28. Freund, K.B.; Laud, K.; Lima, L.H.; Spaide, R.F.; Zweifel, S.; A Yannuzzi, L. Acquired vitelliform lesions. *Retina* **2011**, *31*, 13–25. [CrossRef]

biomedicines

MDPI

Article

Association of *STAT4* Gene Polymorphisms (rs10181656, rs7574865, rs7601754, rs10168266) and Serum STAT4 Levels in Age-Related Macular Degeneration

Tomas Blekeris [1], Greta Gedvilaite [1,2,*], Kriste Kaikaryte [1,2], Loresa Kriauciuniene [2], Dalia Zaliuniene [3] and Rasa Liutkevciene [2,3]

1 Medical Faculty, Medical Academy, Lithuanian University of Health Sciences, LT-50161 Kaunas, Lithuania; tomas.blekeris@stud.lsmu.lt (T.B.); kriste.kaikaryte@lsmuni.lt (K.K.)
2 Neuroscience Institute, Medical Academy, Lithuanian University of Health Sciences, LT-50161 Kaunas, Lithuania; loresa.kriauciuniene@lsmuni.lt (L.K.); rasa.liutkeviciene@lsmuni.lt (R.L.)
3 Ophthalmology Department, Medical Academy, Lithuanian University of Health Sciences, LT-50161 Kaunas, Lithuania; dalia.zaliuniene@lsmuni.lt
* Correspondence: greta.gedvilaite@lsmuni.lt

Abstract: Age-related macular degeneration (AMD) is a progressive degenerative disease that affects the central part of the retina: the macula. AMD is the most common cause of central vision loss in industrialized countries. Increasing attention is being paid to the study of genetic factors that may influence the manifestation of AMD. STAT4 protein is involved in the pathogenesis of numerous inflammatory processes, so we decided to investigate the association between *STAT4* gene polymorphisms (rs10181656, rs7574865, rs7601754, and rs10168266) and age-related macular degeneration. Purpose: To investigate the association between *STAT4* (rs10181656, rs7574865, rs7601754, and rs10168266) gene polymorphisms and STAT4 serum levels in patients with age-related macular degeneration. Methods and participants: The study included 150 individuals with early AMD, 150 individuals with exudative AMD, and 200 healthy subjects. DNA was extracted from peripheral blood leukocytes using the DNA salting-out method, and the genotyping was performed using a real-time polymerase chain reaction (RT-PCR) method. STAT4 serum levels were evaluated using the ELISA method. Statistical analysis was performed using "IBM SPSS "Statistics 29.0" software". Results: The study revealed no statistically significant differences in the distribution of genotypes and alleles for the *STAT4* polymorphisms (rs10181656, rs7574865, rs7601754, and rs10168266) between patients with AMD and the control group. Similarly, a gender-based analysis did not yield any significant differences in the genotype or allele frequencies. Age group comparisons also showed no statistically significant variations in the presence of these *STAT4* polymorphisms between AMD patients and the control group. However, notably, individuals with exudative AMD displayed lower levels of serum STAT4 in comparison to the control group (median (IQR): 0.118 (0.042) vs. 0.262 (0.385), $p = 0.005$). Conclusion: Investigating *STAT4* gene polymorphisms (rs10181656, rs7574865, rs7601754, and rs10168266) did not reveal a significant association with AMD. However, further analysis demonstrated intriguing findings regarding serum STAT4 levels. Exudative AMD patients with at least one G allele of the *STAT4* rs10181656 exhibited significantly lower serum STAT4 levels than the control group subjects ($p = 0.011$). Similarly, those with at least one T allele of STAT4 rs10168266 had lower serum STAT4 levels compared to the control group subjects ($p = 0.039$). These results suggest a potential link between specific *STAT4* genotypes and serum STAT4 levels in exudative AMD patients, shedding light on a novel aspect of the disease.

Keywords: age-related macular degeneration; *STAT4* gene polymorphisms

Citation: Blekeris, T.; Gedvilaite, G.; Kaikaryte, K.; Kriauciuniene, L.; Zaliuniene, D.; Liutkevciene, R. Association of *STAT4* Gene Polymorphisms (rs10181656, rs7574865, rs7601754, rs10168266) and Serum STAT4 Levels in Age-Related Macular Degeneration. *Biomedicines* **2024**, *12*, 18. https://doi.org/10.3390/biomedicines12010018

Academic Editor: Oyuna Kozhevnikova

Received: 13 November 2023
Revised: 12 December 2023
Accepted: 19 December 2023
Published: 20 December 2023

1. Introduction

Age-related macular degeneration (AMD) is a progressive degenerative disease affecting the central part of the retina: the macula. AMD is the most common cause of central vision loss in developed countries. AMD usually occurs in people older than 55 years. As the population ages, AMD is becoming an increasingly important and sensitive disease worldwide [1]. It is estimated that the number of people with AMD will increase from 196 million in 2020 to 288 million in 2040 [2]. The effects of various factors influence the occurrence of AMD. Age has the greatest influence on the development of the disease, but comorbidities, lifestyle, smoking, hypertension, cholesterol, and high BMI are also important [3–5]. Currently, there is no cure for AMD, but it is possible to effectively halt the progression of the disease, and for this early diagnosis of the disease is important. Currently, there is no treatment for dry AMD. Vascular endothelial growth factor inhibitors are the main treatment for exudative AMD [2]. To prevent the onset and progression of the disease, control of modifiable risk factors is important. The pathogenic mechanisms of AMD are not well understood, but it is well established that the development of AMD is influenced by lifestyle, environment, metabolism, and genetic factors [6]. The risk factors for AMD are very similar to the risk factors for cardiovascular disease: age, smoking, high cholesterol, hypertension, and high body mass index [3–5]. The drusen that occur in AMD are composed of the same protein complexes as atherosclerotic plaques, so it is only natural that research be conducted to determine whether cardiovascular disease is associated with the manifestation of AMD [4]. The European Ocular Epidemiology (E3) Consortium conducted the European Eye-Risk Project, which found that elevated high-density lipoprotein cholesterol was associated with AMD risk and larger drusen area. In contrast, higher triglycerides were associated with smaller drusen and lower AMD risk. Diabetes mellitus is also being investigated as a potential risk factor for AMD, as both conditions are associated with increased oxidative stress [3]. The increasing use of smart devices that emit blue light in everyday life has been reported in the literature as a potential risk factor for eye disease [2].

Currently, increasing attention is being paid to the study of genetic factors that may influence the manifestation of AMD. The main pathogenetic mechanisms leading to the development of AMD are the formation of drusen, local inflammation and neovascularization. Stat4, a transcription factor known for its regulatory role in pro-inflammatory signaling, promotes great vessels (GV) vasculogenesis in zebrafish. Some of the Stat4-related pro-inflammatory factors may be involved in large vessel vasculogenesis. Recent studies have revealed a paradigm in which the endogenous mechanisms of pro-inflammatory factors contribute to the maintenance of normal tissue homeostasis [7]. Therefore, pro-inflammatory cytokines and chemokines have also been reported to be expressed in the hearts of infants with congenital heart disease and large vascular defects [8]. But STAT4 is expressed at low levels in cultured human umbilical vein endothelial cells and is tyrosine phosphorylated by interferon [9]. We therefore hypothesized that STA4 may play an important role in the pathogenesis of AMD by influencing the formation of new vessels and wanted to test whether low or high serum STAT4 levels influence the development of AMD.

STAT4 protein is also involved in the pathogenesis of many inflammatory and autoimmune diseases and has been associated with rheumatoid arthritis and systemic lupus erythematosus [10]. Signal Transducers and Activators of Transcription (STAT) are a family of proteins responsible for regulating numerous processes related to cell proliferation, differentiation, apoptosis, and immune response. STAT proteins reside in the cytoplasm and are activated by cytokines and growth factors. After activation, proteins translocate to the nucleus, bind to specific promoters, and regulate gene transcription. The STAT protein family consists of seven members, STAT1, STAT2, STAT3, STAT4, STAT5A, STAT5B, and STAT6, which have been identified in human and mouse genomes. The size of these proteins varies between 750 and 850 amino acids [11].

Each protein of the STAT family plays a different role in signal transduction and is crucial for the cellular response to various cytokines [12]. The *STAT4* gene is located on the long arm of chromosome 2 at position 2q32.2–2q32.3. STAT proteins consist of different regions that differ in structure and function. The N-terminal domain is 124–145 amino acids long. N-terminal and SH2 dimers are mediators of dimerization, allowing the free ends to form STAT dimer complexes and bind to DNA [11]. The SH2 domain is a key mediator of dimerization and a critical factor at the STAT protein–receptor interface. The coiled-coil domain consists of four long α-helices and interacts with other proteins and STAT protein domains [12]. The DNA-binding domain is the C-terminal portion of the protein [13]. The transcription activation domain increases transcriptional activity after serine phosphorylation. The linker domain connects the DNA-binding domain to the SH2 domain [11].

The fourth member of the STAT protein family, the transcriptional signal transducer and activator STAT4, is localized in the cytoplasm. Various cytokines phosphorylate STAT4 after membrane binding and dimerized STAT4 migrates to the nucleus to regulate gene expression [14]. STAT4 is involved in developing many autoimmune and inflammatory diseases and plays a key role in tumor and inflammatory processes. STAT4 protein is crucial for targeting interleukin-12; therefore, IL-12 activates STAT4. The major functions of IL-12 are the production of interferon-ɣ(IFN-ɣ) and the differentiation of Th1 cells into Th17 [15]. Several single nucleotide polymorphisms of the *STAT4* gene have been associated with diseases such as rheumatoid arthritis, Sjogren's, asthma, and systemic lupus erythematosus [10]. In addition to the already known associations of the *STAT4* gene with the aforementioned diseases, it is important to find other diseases whose development may be influenced by the *STAT4* gene. As a result, new studies are being conducted with the *STAT4* gene. In one such study, conducted among the Chinese Han population, it was reported that the *STAT4* gene polymorphisms rs3821236, rs11893432, rs11889341, rs7574865, and rs897200 are associated with the risk of developing type 2 diabetes [16]. *STAT4* gene expression is also associated with the risk of type 1 diabetes. A study conducted in Poland found an association between the rs7574865 polymorphism of the *STAT4* gene and the risk of type 1 diabetes in a population of Polish European children [17]. The study included 1438 individuals whose genotypes were compared, including 656 children with type 1 diabetes and 782 healthy adults as a control group. According to scientific research, *STAT4* gene expression is associated with the risk of both type 1 and type 2 diabetes [16,17]. In a study conducted in western China involving 725 individuals, *STAT4* polymorphisms rs7574865, rs10181656, rs10168266, and rs13426947 were also found to be associated with the risk of neuromyelitis optica spectrum disorder [18]. Our study uniquely integrated the analysis of single nucleotide polymorphisms (SNPs) with serum STAT4 levels in blood serum, providing a comprehensive approach to understanding AMD. This dual analysis considers both genetic predispositions, particularly in genes like *STAT4*, and systemic factors reflected in serum biomarkers. By exploring this interplay, our research aims to contribute novel insights into AMD pathogenesis, potentially informing diagnostic and therapeutic strategies. To discover new genetic markers associated with the development of AMD, we decided to investigate the single nucleotide polymorphisms rs10181656, rs7574865, rs7601754, and rs10168266 of the *STAT4* gene and determine their influence on the manifestation of AMD.

2. Methods

All subjects signed an agreement in accordance with the Declaration of Helsinki. The study was conducted in the Laboratory of Ophthalmology of the Institute of Neurosciences of the Lithuanian University of Health Sciences. The Kaunas Regional Biomedical Research Ethics Committee approved the study (approval numbers: 9 July 2015 No. BE-2-26 and 26 January 2017 No. P1-BE-2-26/2015).

A total of 500 subjects were studied, and two study groups were formed: a control group (n = 200) and a group of patients with AMD (n = 300). The patient group was divided into two subgroups: patients with early AMD (n = 150) and patients with exudative AMD (n = 150).

The control group consisted of individuals who had no ocular pathology at examination and agreed to participate in the study. The exclusion criteria for patients in the study were described in our previous study.

The exclusion criteria for patients with AMD were (1) related ocular diseases (high refractive error, cloudy cornea, or lens opacity (nuclear, cortical, and posterior subcapsular cataract), excluding minor opacities, and patients with intraocular lenses, keratitis, acute or chronic uveitis, glaucoma, late age-related macular degeneration, optic nerve disease); (2) systemic diseases (diabetes mellitus, oncological diseases, systemic tissue disorders, chronic infectious diseases, and conditions after organ or tissue transplantation); (3) color fundus photography because of the opacity of the optical system of the eye or because of the quality of fundus photography; (4) congenital color vision disorders were excluded by history; and (5) patients with epilepsy and taking sedatives.

The inclusion criteria for healthy patients were as follows: (1) no ophthalmic eye diseases detected in the detailed ophthalmic examination; (2) informed consent to participate. The exclusion criteria for healthy patients were as follows: (1) any ophthalmic diseases; (2) patients with epilepsy and taking sedatives.

Ophthalmic examination of all subjects in our study was performed as follows. Visual acuity (VA) was estimated from letter charts and reported in decimal notation. All patients were examined with slit-lamp biomicroscopy. Biomicroscopy was used to assess corneal and lens transparency. Intraocular pressure was measured at each examination. The patients were dilated with 1% tropicamide. After pupil dilation, funduscopy was performed with a direct monocular ophthalmoscope and slit lamp using a double aspheric lens of +78 diopters. The examination results were recorded on standardized forms developed for this study. Color fundus photographs were taken with a fundus camera at half wide angle (OPTON SBG, 30 degrees). Photographs were taken with the focus on the center of the fovea.

Optical coherence tomography was performed in all AMD patients (OCT), and fluorescein angiography was performed in patients with suspected late AMD after examination of the OCT. For this study, we used the classification system for AMD formulated in the previous Age-Related Eye Disease Study [6,19]: Early AMD consisted of multiple small drusen and multiple intermediate (63–124 μm diameter) drusen or abnormalities of the retinal pigment epithelium. Extensive intermediate drusen characterized early intermediate AMD and at least one large druse ($\geq$125 μm diameter) or geographic atrophy that did not involve the center of the fovea. Exudative AMD was identified by the occurrence of geographic atrophy involving the fovea and/or any of the neovascular AMD features.

The control group consisted of individuals who had no ocular pathology at examination and agreed to participate in the study.

2.1. DNA Extraction and Genotyping

The DNA extraction and analysis of the *STAT4* gene polymorphisms were performed at the Ophthalmology Laboratory of the Neuroscience Institute of the Lithuanian College of Health Sciences. Blood samples were typically collected before 10 am. After obtaining the blood, the sample was immediately delivered to the laboratory. The sample was coded and labeled according to the laboratory's instructions. Then, it was either used for DNA extraction or refrigerated for future use. For serum preparation, upon collection of the whole blood, it was left undisturbed at room temperature for 15–30 min to allow for clotting. Subsequently, the clot was removed by centrifuging the blood at 1000–2000$\times$ *g* for 10 min in a refrigerated centrifuge, resulting in the designated serum (i.e., supernatant). This serum was immediately transferred into a clean microcentrifuge tube using a pipette. Throughout handling, the samples were maintained at 2–8 °C. If the serum was not analyzed immediately, it was stored and transported at −20 °C or lower for further investigations. Approximately 3 mL of blood is required to extract approximately 250 μg of DNA, a quantity optimal for various tests, including TL-PCR. The DNA extraction was performed on the venous blood samples using the salting-out method. Briefly, venous blood samples (i.e., white blood cells) were collected and suspended in a buffer solution, followed by the addition of detergents to degrade cell membranes, proteinase K to hydrolyze proteins, and chloroform to deproteinize them. The DNA was then precipitated with ethanol. Additionally, the DNA concentration was determined using spectrophotometry. The 260/280, 260/230, and 260/325 absorbance ratios were used in assessing the DNA purity and identifying contaminants in the biological samples during DNA extraction. To ensure optimal accuracy, the readings ideally fell within the range of 0.1 to 1.0.

Single nucleotide polymorphisms (SNPs) were determined using TaqMan® genotyping assays (Thermo Scientific, Pleasanton, CA, USA, Canada) and following the manufacturer's. Genotyping of *STAT4* rs10181656, rs7574865, rs7601754, and rs10168266 was performed using real-time PCR (RT-PCR) according to the manufacturers, using a Step One Plus RT-PCR system (Applied Biosystems, Foster City, CA, USA) and an allele discrimination program. Into each of the 96 wells on the plate, we added 1.5 μL of the DNA samples and 8.5 μL of the PCR reaction mixture, along with the negative control. The program analyzed each genotype based on the fluorescence intensity of the different detectors (VIC and FAM).

2.2. ELISA

STAT4 levels were determined using an enzyme-linked immunosorbent assay (ELISA) with the Abbexa Signal Transducer And Activator Of Transcription 4 (STAT4) ELISA kit (UK, Cambridge).

In this method, a 96-well plate was precoated with an antibody. Following the addition of standards, test samples, and a biotin-conjugated reagent, the plate was incubated. Subsequently, an HRP-conjugated reagent was introduced, and the plate underwent another incubation. Wash buffer was used to remove any unbound conjugates at each stage. The HRP enzymatic reaction was quantified using a TMB substrate, resulting in a blue-colored product in the wells containing adequate STAT4, which transformed to yellow upon the addition of an acidic stop solution. The intensity of the yellow color was directly proportional to the amount of STAT4 bound to the plate. The optical density (OD) was measured at 450 nm using a microplate reader, enabling the calculation of the concentration of STAT4. The absorbance was measured at the required 450 nm, and the concentration was calculated from a calibration curve based on the standard solutions used.

2.3. Statistical Analysis

Statistical analysis of the data was performed using IBM SPSS Statistics 27.0. Data are presented as absolute numbers (percentages) and the median (IQR). The Mann–Whitney U test was used to detect differences between two independent groups. To compare the homogeneity of the genotype distribution of polymorphisms between AMD patients and controls, $\chi 2$ and Fisher's and two-way criteria were used. Binary logistic regression analysis was used to estimate the odds ratio (OR) of AMD occurrence as a function of genetic inheritance patterns. The genetic models (codominant: heterozygotes vs. wild-type homozygotes and homozygotes vs. wild-type homozygotes; dominant: homozygotes with a rarer allele and heterozygotes vs. wild-type homozygotes; recessive: homozygotes with a rarer allele vs. wild-type homozygotes; recessive: homozygotes with a rarer allele vs. wild-type homozygotes) were included in the analysis. Homozygotes with rarer allele vs wild-type homozygotes and heterozygotes; supradominant: heterozygotes vs wild-type homozygotes vs homozygotes with rarer allele; an additive model was used to model the effect of each rarer allele on the development of AMD. This analysis was performed with a 95% confidence interval (CI) for the group with AMD. The Akaike information criterion (AIC) was evaluated to select the best inheritance model, with the lowest value indicating the best-fitting model. After Bonferroni correction, differences were considered statistically significant when $p < 0.05/4$ ($p < 0.0125$).

3. Results

This case–control study included 500 participants: 150 subjects in the early AMD group (average age: 71.49 years) and 150 subjects in the exudative AMD group (average age: 71.46 years). The control group comprised 200 healthy subjects (average age: 71.42 years) (Table 1).

Table 1. Demographic characteristics.

Characteristics		Group			*p*-Value
		Early AMD $n = 150$	Exudative AMD $n = 150$	Control $n = 200$	
Gender	Women, N (%)	75 (50)	75 (50)	100 (50)	1 *
	Men, N (%)	75 (50)	75 (50)	100 (50)	1 **
Interquartile range (IQR)		71 (11)	72.5 (11)	71 (4)	0.726 * 0.152 **

AMD—age-related macular degeneration. * Early AMD vs. control group. ** Exudative AMD vs. control group.

3.1. Frequencies of STAT4 (rs10181656, rs7574865, rs7601754, and rs10168266) Genotypes and Alleles in Patients with AMD and Control Group

No statistically significant differences were found between the frequencies of *STAT4* (rs10181656, rs7574865, rs7601754, and rs10168266) genotypes and alleles in early AMD and control groups (Table 2).

Table 2. Frequencies of *STAT4* (rs10181656, rs7574865, rs7601754, and rs10168266) genotypes and alleles in the early AMD and control groups.

SNP	Genotype/Allele	Group		*p*-Value
		Control, N (%)	Early AMD, N (%)	
STAT4 rs10181656	CC	118 (59.0)	90 (60.0)	0.672
	CG	65 (32.5)	51 (34.0)	
	GG	17 (8.5)	9 (6.0)	
	Total:	200	150	
	Allele:			
	C	301 (75.25)	231 (77.0)	
	G	99 (24.75)	69 (23.0)	0.592
STAT4 rs7574865	GG	118 (59.0)	92 (61.3)	0.670
	GT	65 (32.5)	49 32.7)	
	TT	17 (8.5)	9 (6.0)	
	Total:	200	150	
	Allele:			
	G	301 (75.25)	233 (77.67)	
	T	99 (24.75)	67 (22.33)	0.457
STAT4 rs7601754	AA	150 (75.0)	113 (75.3)	0.902
	GA	46 (23.0)	33 (22.0)	
	GG	4 (2.0)	4 (2.7)	
	Total:	200	150	
	Allele:			
	A	346 (86.5)	259 (86.34)	
	G	54 (13.5)	41 (13.66)	0.949
STAT4 rs10168266	CC	133 (66.5)	100 (66.7)	0.852
	CT	58 (29.0)	45 (30.0)	
	TT	9 (4.5)	5 (3.3)	
	Total:	200	150	
	Allele:			
	C	324 (81.0)	245 (81.67)	
	T	76 (19.0)	55 (18.33)	0.823

p—Significance level and Bonferroni-corrected significance level when $p < 0.05/4$; SNP—single nucleotide polymorphism; AMD—age-related macular degeneration.

Moreover, no statistically significant differences were found between the frequencies of the *STAT4* (rs10181656, rs7574865, rs7601754, and rs10168266) genotypes and alleles in the exudative AMD and control groups (Table 3).

Table 3. Frequencies of the *STAT4* (rs10181656, rs7574865, rs7601754, and rs10168266) genotypes and alleles in the exudative AMD and control groups.

SNP	Genotype/Allele	Group		p-Value
		Control, N (%)	Exudative AMD, N (%)	
STAT4 rs10181656	CC	118 (59.0)	80 (53.3)	0.568
	CG	65 (32.5)	56 (37.3)	
	GG	17 (8.5)	14 (9.3)	
	Total:	200	150	
	Allele:			
	C	301 (75.25)	216 (72.0)	
	G	99 (24.75)	84 (28.0)	0.333
STAT4 rs7574865	GG	118 (59.0)	80 (53.3)	0.568
	GT	65 (32.5)	56 (37.3)	
	TT	17 (8.5)	14 (9.3)	
	Total:	200	150	
	Allele:			
	G	301 (75.25)	216 (72.0)	
	T	99 (24.75)	84 (28.0)	0.333
STAT4 rs7601754	AA	150 (75.0)	111 (74.0)	0.529
	GA	46 (23.0)	38 (25.3)	
	GG	4 (2.0)	1 (0.7)	
	Total:	200	150	
	Allele:			
	A	346 (86.5)	260 (86.67)	
	G	54 (13.5)	40 (13.33)	0.949
STAT4 rs10168266	CC	133 (66.5)	98 (65.3)	0.674
	CT	58 (29.0)	42 (28.0)	
	TT	9 (4.5)	10 (6.7)	
	Total:	200	150	
	Allele:			
	C	324 (81.0)	238 (79.34)	
	T	76 (19.0)	62 (20.96)	0.583

p—Significance level and Bonferroni-corrected significance level when $p < 0.05/4$; SNP—single nucleotide polymorphism; AMD—age-related macular degeneration.

3.2. *STAT4 (rs10181656, rs7574865, rs7601754, and rs10168266) Genotypes and Allele Associations with Early and Exudative AMD*

We analyzed *STAT4* (rs10181656, rs7574865, rs7601754, and rs10168266) genotypes and allele associations with early and exudative AMD. No statistically significant associations were found (Tables 4 and 5).

Table 4. Binomial logistic regression analysis of *STAT4* (rs10181656, rs7574865, rs7601754, and rs10168266) in the early AMD and control groups.

Model	Genotype/Allele	OR (95% CI)	p-Value	AIC
		STAT4 rs10181656		
Codominant	CG vs. CC	1.029 (0.651–1.626)	0.904	481.227
	GG vs. CC	0.694 (0.296–1.629)	0.402	
Dominant	CG+GG vs. CC	0.959 (0.623–1.477)	0.850	480.000
Recessive	GG vs. CC+CG	0.687 (0.297–1.587)	0.380	479.241
Overdominant	CG vs. CC+GG	1.070 (0.683–1.676)	0.768	479.949
Additive	C	0.915 (0.653–1.283)	0.608	479.771

Table 4. *Cont.*

Model	Genotype/Allele	OR (95% CI)	*p*-Value	AIC
		STAT4 rs7574865		
Codominant	GT vs. GG TT vs. GG	0.967 (0.610–1.532) 0.679 (0.289–1.593)	0.886 0.374	481.221
Dominant	GT+TT vs. GG	0.907 (0.588–1.399)	0.659	479.841
Recessive	TT vs. GG+GT	0.687 (0.297–1.587)	0.380	479.241
Overdominant	GT vs. GG+TT	1.008 (0.641–1.583)	0.974	480.035
Additive	G	0.885 (0.631–1.241)	0.478	479.530
		STAT4 rs7601754		
Codominant	GA vs. AA GG vs. AA	0.952 (0.572–1.585) 1.327 (0.325–5.422)	0.851 0.693	481.831
Dominant	GA+GG vs. AA	0.982 (0.602–1.604)	0.943	480.031
Recessive	GG vs. AA+GA	1.342 (0.330–5.457)	0.681	479.867
Overdominant	GA vs. AA+GG	0.944 (0.568–1.569)	0.825	479.987
Additive	A	1.014 (0.660–1.556)	0.950	480.032
		STAT4 rs10168266		
Codominant	CT vs. CC TT vs. CC	1.032 (0.646–1.648) 0.739 (0.240–2.273)	0.895 0.598	481.709
Dominant	CT+TT vs. CC	0.993 (0.634–1.555)	0.974	480.035
Recessive	TT vs. CC+CT	0.732 (0.240–2.230)	0.583	479.727
Overdominant	CT vs. CC+TT	1.049 (0.660–1.669)	0.839	479.994
Additive	C	0.958 (0.656–1.399)	0.826	479.987

p—Significance level and Bonferroni-corrected significance level when $p < 0.05/4$; OR—odds ratio; CI—confident interval; AIC—Akaike information criteria.

Table 5. Binomial logistic regression analysis of *STAT4* (rs10181656, rs7574865, rs7601754, and rs10168266) in the exudative AMD and control groups.

Model	Genotype/Allele	OR (95% CI)	*p*-Value	AIC
		STAT4 rs10181656		
Codominant	CG vs. CC GG vs. CC	1.271 (0.805–2.006) 1.215 (0.567–2.603)	0.303 0.617	480.904
Dominant	CG+GG vs. CC	1.259 (0.822–1.930)	0.290	478.916
Recessive	GG vs. CC+CG	1.108 (0.528–2.326)	0.786	479.962
Overdominant	CG vs. CC+GG	1.237 (0.794–1.929)	0.347	479.153
Additive	C	1.164 (0.842–1.608)	0.357	479.189
		STAT4 rs7574865		
Codominant	GT vs. GG TT vs. GG	1.271 (0.805–2.006) 1.215 (0.567–2.603)	0.303 0.617	480.904
Dominant	GT+TT vs. GG	1.259 (0.822–1.930)	0.290	478.916
Recessive	TT vs. GG+GT	1.108 (0.528–2.326)	0.786	479.962
Overdominant	GT vs. GG+TT	1.237 (0.794–1.929)	0.347	479.153
Additive	G	1.164 (0.842–1.608)	0.357	479.189

Table 5. *Cont.*

Model	Genotype/Allele	OR (95% CI)	*p*-Value	AIC
		STAT4 rs7601754		
Codominant	GA vs. AA GG vs. AA	1.116 (0.681–1.831) 0.338 (0.037–3.064)	0.663 0.335	480.663
Dominant	GA+GG vs. AA	1.054 (0.649–1.713)	0.832	479.991
Recessive	GG vs. AA+GA	0.329 (0.036–2.973)	0.322	478.853
Overdominant	GA vs. AA+GG	1.136 (0.693–1.861)	0.613	479.781
Additive	A	0.985 (0.630–1.540)	0.948	480.031
		STAT4 rs10168266		
Codominant	CT vs. CC TT vs. CC	0.983 (0.611–1.581) 1.508 (0.590–3.851)	0.943 0.391	481.256
Dominant	CT+TT vs. CC	1.053 (0.674–1.646)	0.820	479.984
Recessive	TT vs. CC+CT	1.516 (0.600–3.829)	0.379	479.261
Overdominant	CT vs. CC+TT	0.952 (0.595–1.522)	0.838	479.994
Additive	C	1.100 (0.769–1.574)	0.601	479.762

p—Significance level and Bonferroni-corrected significance level when $p < 0.05/4$; OR—odds ratio; CI—confident interval; AIC—Akaike information criteria.

3.3. *STAT4 (rs10181656, rs7574865, rs7601754, and rs10168266) Genotypes and Allele Associations with Early and Exudative AMD by Gender*

The allele frequency analysis showed no statistically significant differences between *STAT4* (rs10181656, rs7574865, rs7601754, and rs10168266) genotypes and alleles in the early AMD, exudative AMD, and control groups depending on gender (Tables 6 and 7).

Table 6. Frequencies of *STAT4* (rs10181656, rs7574865, rs7601754, and rs10168266) genotypes and alleles in the early AMD and control groups by gender.

SNP	Genotype/Allele	Women		*p*-Value	Men		*p*-Value
		Control, N (%)	Early AMD, N (%)		Control, N (%)	Early AMD, N (%)	
STAT4 rs10181656	CC	58 (58.0)	48 (64.0)	0.694	60 (60.0)	42 (56.0)	0.574
	CG	33 (33.0)	22 (29.3)		32 (32.0)	29 (38.7)	
	GG	9 (9.0)	5 (6.7)		8 (8.0)	4 (5.3)	
	Allele:						
	C	149 (74.5)	118 (78.67)	0.364	152 (76.0)	113 (75.34)	0.886
	G	51 (25.5)	32 (21.33)		48 (24.0)	37 (24.66)	
STAT4 rs7574865	GG	58 (58.0)	48 (64.0)	0.722	60 (60.0)	44 (58.7)	0.482
	GT	33 (33.0)	21 (28.0)		32 (32.0)	28 (37.3)	
	TT	9 (9.0)	6 (8.0)		8 (8.0)	3 (4.0)	
	Allele:						
	G	149 (74.5)	117 (78.0)	0.448	152 (76.0)	116 (77.34)	0.771
	T	51 (25.5)	33 (22.0)		48 (24.0)	34 (22.66)	
STAT4 rs7601754	AA	73 (73.0)	50 (66.7)	0.411	77 (77.0)	63 (84.0)	0.312
	GAGG	25 (25.0)	21 (28.0)		21 (21.0)	12 (16.0)	
		2 (2.0)	4 (5.3)		2 (2.0)	0 (0.0)	
	Allele:						
	A	171 (85.5)	121 (80.67)	0.229	175 (87.5)	138 (92.0)	0.175
	G	29 (14.5)	29 (19.33)		25 (12.5)	12 (8.0)	

Table 6. *Cont.*

SNP	Genotype/Allele	Women		p-Value	Men		p-Value
		Control, N (%)	Early AMD, N (%)		Control, N (%)	Early AMD, N (%)	
STAT4 rs10168266	CC	62 (62.0)	50 (66.7)	0.749	71 (71.0)	50 (66.7)	0.792
	CT	32 (32.0)	22 (29.3)		26 (26.0)	23 (30.7)	
	TT	6 (6.0)	3 (4.0)		3 (3.0)	2 (2.7)	
	Allele:						
	C	156 (78.0)	122 (81.33)	0.445	168 (84.0)	123 (82.0)	0.621
	T	44 (22.0)	28 (18.67)		32 (16.0)	27 (18.0)	

p—Significance level and Bonferroni-corrected significance level when $p < 0.05/4$; SNP—single nucleotide polymorphism; AMD—age-related macular degeneration.

Table 7. Frequencies of *STAT4* (rs10181656, rs7574865, rs7601754, and rs10168266) genotypes and alleles in the exudative AMD and control groups by gender.

SNP	Genotype/Allele	Women		p-Value	Men		p-Value
		Control, N (%)	Exudative AMD, N (%)		Control, N (%)	Exudative AMD, N (%)	
STAT4 rs10181656	CC	58 (58.0)	37 (49.3)	0.494	60 (60.0)	43 (57.3)	0.921
	CG	33 (33.0)	31 (41.3)		32 (32.0)	25 (33.3)	
	GG	9 (9.0)	7 (9.3)		8 (8.0)	7 (9.3)	
	Allele:						
	C	149 (74.5)	105 (70.0)	0.350	152 (76.0)	111 (74.0)	0.668
	G	51 (25.5)	45 (30.0)		48 (24.0)	39 (26.0)	
STAT4 rs7574865	GG	58 (58.0)	37 (49.3)	0.494	60 (60.0)	43 (57.3)	0.921
	GT	33 (33.0)	31 (41.3)		32 (32.0)	25 (33.3)	
	TT	9 (9.0)	7 (9.3)		8 (8.0)	7 (9.3)	
	Allele:						
	G	149 (74.5)	105 (70.0)	0.350	152 (76.0)	111 (74.0)	0.668
	T	51 (25.5)	45 (30.0)		48 (24.0)	39 (26.0)	
STAT4 rs7601754	AA	73 (73.0)	54 (72.0)	0.438	77 (77.0)	57 (76.0)	0.918
	GA	25 (25.0)	21 (28.0)		21 (21.0)	17 (22.7)	
	GG	2 (2.0)	0 (0)		2 (2.0)	1 (1.3)	
	Allele:						
	A	171 (85.5)	129 (86.0)	0.895	175 (87.5)	131 (87.34)	0.963
	G	29 (14.5)	21 (14.0)		25 (12.5)	19 (12.66)	
STAT4 rs10168266	CC	62 (62.0)	47 (62.7)	0.972	71 (71.0)	51 (68.0)	0.516
	CT	32 (32.0)	23 (30.7)		26 (26.0)	19 (25.3)	
	TT	6 (6.0)	5 (6.7)		3 (3.0)	5 (6.7)	
	Allele:						
	C	156 (78.0)	117 (78.0)	1	168 (84.0)	121 (80.67)	0.416
	T	44 (22.0)	33 (22.0)		32 (16.0)	29 (19.33)	

p—Significance level and Bonferroni-corrected significance level when $p < 0.05/4$; SNP—single nucleotide polymorphism; AMD—age-related macular degeneration.

Binary logistic regression revealed that the association between *STAT4* (rs10181656, rs7574865, rs7601754, and rs10168266) in the early AMD and control groups by gender was not statistically significant (Table 8). Also, no statistically significant association was found between *STAT4* (rs10181656, rs7574865, rs7601754, and rs10168266) in the exudative AMD and control groups, depending on gender (Table 9).

Table 8. Binary logistic regression analysis of *STAT4* (rs10181656, rs7574865, rs7601754, and rs10168266) in the early AMD and control groups by gender.

Model	Genotype/Allele	OR (95% CI)	*p*-Value	AIC
		Women		
		STAT4 rs10181656		
Codominant	CG vs. CC GG vs. CC	0.806 (0.416–1.561) 0.671 (0.211–2.137)	0.522 0.500	242.283
Dominant	CG+GG vs. CC	0.777 (0.419–1.439)	0.422	240.370
Recessive	GG vs. CC+CG	0.722 (0.232–2.251)	0.575	240.696
Overdominant	CG vs. CC+GG	0.843 (0.441–1.612)	0.605	240.750
Additive	C	0.814 (0.506–1.309)	0.395	240.285
		STAT4 rs7574865		
Codominant	GT vs. GG TT vs. GG	0.769 (0.394–1.499) 0.806 (0.268–2.424)	0.440 0.700	242.364
Dominant	GT+TT vs. GG	0.777 (0.419–1.439)	0.422	240.370
Recessive	TT vs. GG+GT	0.879 (0.299–2.587)	0.815	240.963
Overdominant	GT vs. GG+TT	0.790 (0.411–1.519)	0.479	240.513
Additive	G	0.845 (0.530–1.348)	0.481	240.516
		STAT4 rs7601754		
Codominant	GA vs. AA GG vs. AA	1.226 (0.620–2.427) 2.920 (0.515–16.555)	0.558 0.226	239.247
Dominant	GA+GG vs. AA	1.352 (0.704–2.595)	0.365	240.198
Recessive	GG vs. AA+GA	2.761 (0.492–15.488)	0.249	239.590
Overdominant	GA vs. AA+GG	1.167 (0.593–2.297)	0.656	240.819
Additive	A	1.392 (0.799–2.424)	0.242	239.646
		STAT4 rs10168266		
Codominant	CT vs. CC TT vs. CC	0.853 (0.441–1.647) 0.620 (0.148–2.604)	0.635 0.514	242.431
Dominant	CT+TT vs. CC	0.816 (0.436–1.528)	0.525	240.611
Recessive	TT vs. CC+CT	0.653 (0.158–2.699)	0.556	240.658
Overdominant	CT vs. CC+TT	0.882 (0.460–1.691)	0.706	240.875
Additive	C	0.822 (0.490–1.379)	0.458	240.461
		Men		
		STAT4 rs10181656		
Codominant	CG vs. CC GG vs. CC	1.295 (0.684–2.452) 0.714 (0.202–2.527)	0.428 0.602	241.902
Dominant	CG+GG vs. CC	1.179 (0.643–2.162)	0.595	240.736
Recessive	GG vs. CC+CG	0.648 (0.188–2.238)	0.493	240.529
Overdominant	CG vs. CC+GG	1.340 (0.716–2.507)	0.360	240.182
Additive	C	1.035 (0.640–1.674)	0.888	240.998
		STAT4 rs7574865		
Codominant	GT vs. GG TT vs. GG	1.193 (0.630–2.261) 0.511 (0.128–2.038)	0.588 0.342	241.505
Dominant	GT+TT vs. GG	1.057 (0.575–1.944)	0.859	240.986

Table 8. *Cont.*

Model	Genotype/Allele	OR (95% CI)	*p*-Value	AIC
Recessive	TT vs. GG+GT	0.479 (0.123–1.871)	0.290	239.798
Overdominant	GT vs. GG+TT	1.266 (0.675–2.374)	0.462	240.478
Additive	G	0.931 (0.570–1.521)	0.776	240.936
		STAT4 rs7601754		
Codominant	GA vs. AA GG vs. AA	0.698 (0.319–1.529) -	0.369 -	239.941
Dominant	GA+GG vs. AA	0.638 (0.294–1.382)	0.254	239.683
Recessive	GG vs. AA+GA	-	-	-
Overdominant	GA vs. AA+GG	0.717 (0.328–1.567)	0.404	240.309
Additive	A	0.608 (0.294–1.259)	0.180	239.135
		STAT4 rs10168266		
Codominant	CT vs. CC TT vs. CC	1.256 (0.644–2.449) 0.947 (0.153–5.874)	0.503 0.953	242.553
Dominant	CT+TT vs. CC	1.224 (0.642–2.335)	0.539	240.642
Recessive	TT vs. CC+CT	0.886 (0.144–5.439)	0.896	241.001
Overdominant	CT vs. CC+TT	1.259 (0.648–2.445)	0.497	240.557
Additive	C	1.152 (0.656–2.023)	0.621	240.774

p—Significance level and Bonferroni-corrected significance level when $p < 0.05/4$; OR—odds ratio; CI—confident interval; AIC—Akaike information criteria.

Table 9. Binomial logistic regression analysis of *STAT4* (rs10181656, rs7574865, rs7601754, and rs10168266) in the exudative AMD and control groups by gender.

Model	Genotype/Allele	OR (95% CI)	*p*-Value	AIC
		Women		
		STAT4 rs10181656		
Codominant	CG vs. CC GG vs. CC	1.473 (0.776–2.794) 1.219 (0.418–3.556)	0.236 0.717	241.608
Dominant	CG+GG vs. CC	1.418 (0.777–2.590)	0.255	239.721
Recessive	GG vs. CC+CG	1.041 (0.369–2.934)	0.940	241.012
Overdominant	CG vs. CC+GG	1.430 (0.769–2.660)	0.258	239.739
Additive	C	1.232 (0.781–1.942)	0.370	240.214
		STAT4 rs7574865		
Codominant	GT vs. GG TT vs. GG	1.473 (0.776–2.794) 1.219 (0.418–3.556)	0.236 0.717	241.608
Dominant	GT+TT vs. GG	1.418 (0.777–2.590)	0.255	239.721
Recessive	TT vs. GG+GT	1.041 (0.369–2.934)	0.940	241.012
Overdominant	GT vs. GG+TT	1.430 (0.769–2.660)	0.258	239.739
Additive	G	1.232 (0.781–1.942)	0.370	240.214
		STAT4 rs7601754		
Codominant	GA vs. AA GG vs. AA	1.136 (0.576–2.238) -	0.713 -	240.627

Table 9. *Cont.*

Model	Genotype/Allele	OR (95% CI)	*p*-Value	AIC
Dominant	GA+GG vs. AA	1.051 (0.538–2.055)	0.883	240.996
Recessive	GG vs. AA+GA	-	-	-
Overdominant	GA vs. AA+GG	1.167 (0.593–2.297)	0.656	240.819
Additive	A	0.957 (0.510–1.796)	0.891	240.999
		STAT4 rs10168266		
Codominant	CT vs. CC TT vs. CC	0.948 (0.492–1.828) 1.099 (0.316–3.821)	0.874 0.882	242.960
Dominant	CT+TT vs. CC	0.972 (0.524–1.803)	0.928	241.010
Recessive	TT vs. CC+CT	1.119 (0.328–3.815)	0.857	240.986
Overdominant	CT vs. CC+TT	0.940 (0.493–1.793)	0.851	240.982
Additive	C	1.000 (0.612–1.634)	1.000	241.018
		Men		
		STAT4 rs10181656		
Codominant	CG vs. CC GG vs. CC	1.090 (0.567–2.095) 1.221 (0.412–3.622)	0.796 0.719	242.854
Dominant	CG+GG vs. CC	1.116 (0.608–2.050)	0.723	240.892
Recessive	GG vs. CC+CG	1.184 (0.409–3.423)	0.755	240.921
Overdominant	CG vs. CC+GG	1.062 (0.561–2.011)	0.852	240.983
Additive	C	1.099 (0.694–1.741)	0.687	240.856
		STAT4 rs7574865		
Codominant	GT vs. GG TT vs. GG	1.090 (0.567–2.095) 1.221 (0.412–3.622)	0.796 0.719	242.854
Dominant	GT+TT vs. GG	1.116 (0.608–2.050)	0.723	240.892
Recessive	TT vs. GG+GT	1.184 (0.409–3.423)	0.755	240.921
Overdominant	GT vs. GG+TT	1.062 (0.561–2.011)	0.852	240.983
Additive	G	1.099 (0.694–1.741)	0.687	240.856
		STAT4 rs7601754		
Codominant	GA vs. AA GG vs. AA	1.094 (0.529–2.259) 0.675 (0.060–7.632)	0.809 0.751	242.844
Dominant	GA+GG vs. AA	1.057 (0.522–2.141)	0.877	240.994
Recessive	GG vs. AA+GA	0.662 (0.059–7.442)	0.738	240.902
Overdominant	GA vs. AA+GG	1.103 (0.535–2.274)	0.791	240.948
Additive	A	1.015 (0.538–1.914)	0.963	241.016
		STAT4 rs10168266		
Codominant	CT vs. CC TT vs. CC	1.017 (0.509–2.033) 2.320 (0.530–10.151)	0.961 0.264	241.709
Dominant	CT+TT vs. CC	1.152 (0.602–2.206)	0.669	240.836
Recessive	TT vs. CC+CT	2.310 (0.534–9.984)	0.262	239.712
Overdominant	CT vs. CC+TT	0.966 (0.486–1.917)	0.920	241.008
Additive	C	1.231 (0.726–2.087)	0.440	240.423

p—Significance level and Bonferroni-corrected significance level when $p < 0.05/4$; OR—odds ratio; CI—confident interval; AIC—Akaike information criteria.

3.4. Frequencies of *STAT4* (rs10181656, rs7574865, rs7601754, and rs10168266) Genotypes and Alleles in Patients with AMD and in the Control Group by Age

Since AMD is a major cause of central vision loss in the developed world affecting 10% of people older than 65 years and more than 25% of people older than 75 years, we decided to divide our subjects into three groups, depending on age (Table 10). Our results showed that there were no statistically significant differences in *STAT4* (rs10181656, rs7574865, rs7601754, and rs10168266) genotypes and alleles among the early AMD, exudative AMD, and control groups, depending on age (Table 10).

Table 10. Frequencies of *STAT4* (rs10181656, rs7574865, rs7601754, and rs10168266) genotypes and alleles in the early AMD, exudative AMD, and control groups by age.

SNP	Genotype/Allele	≤65 y/o				>65 y/o–≤75 y/o				>75 y/o			
		Control Group, n (%)	Early AMD n (%)	Exudative AMD n (%)	*p*-Value	Control Group, n (%)	Early AMD n (%)	Exudative AMD n (%)	*p*-Value	Control Group, n (%)	Early AMD n (%)	Exudative AMD n (%)	*p*-Value
rs10181656	CC	13 (52.0)	15 (50.0)	17 (54.8)	0.965 (1) 0.754 (2)	89 (59.7)	48 (62.3)	32 (50.0)	0.493 (1) 0.411 (2)	16 (61.5)	27 (62.8)	31 (56.4)	0.797 (1) 0.447 (2)
	CG	10 (40.0)	12 (40.0)	10 (32.3)		48 (32.2)	26 (33.8)	25 (39.1)		7 (26.9)	13 (30.2)	21 (38.2)	
	GG	2 (8.0)	3 (10.0)	4 (12.9)		12 (8.1)	3 (3.9)	7 (10.9)		3 (11.5)	3 (7.0)	3 (5.5)	
	C	36 (72)	42 (70)	44 (71)	0.818 (1) 0.904 (2)	226 (75.8)	122 (79.2)	89 (69.5)	0.418 (1) 0.174 (2)	39 (75)	67 (77.9)	83 (75.5)	0.695 (1) 0.950 (2)
	G	14 (28)	18 (30)	18 (29)		72 (24.2)	32 (20.8)	39 (30.5)		13 (25)	19 (22.1)	27 (24.5)	
rs7574865	GG	13 (52.0)	17 (56.7)	17 (54.8)	0.938 (1) 0.754 (2)	89 (59.7)	48 (62.3)	31 (48.4)	0.725 (1) 0.277 (2)	16 (61.5)	27 (62.8)	32 (58.2)	0.797 (1) 0.291 (2)
	GT	10 (40.0)	11 (36.7)	10 (32.3)		48 (32.2)	25 (32.5)	25 (39.1)		7 (26.0)	13 (30.2)	21 (38.2)	
	TT	2 (8.0)	2 (7.7)	4 (12.9)		12 (8.1)	4 (5.2)	8 (12.5)		3 (11.5)	3 (7.0)	2 (3.6)	
	G	36 (72)	45 (75)	44 (71)	0.722 (1) 0.904 (2)	226 (75.8)	121 (78.6)	87 (68)	0.514 (1) 0.092 (2)	39 (75)	67 (77.9)	85 (77.3)	0.695 (1) 0.749 (2)
	T	14 (28)	15 (25)	18 (29)		72 (24.2)	33 (21.4)	41 (32)		13 (25)	19 (22.1)	25 (22.7)	
rs7601754	AA	19 (76.0)	18 (60.0)	21 (67.7)	0.453 (1) 0.342 (2)	113 (75.8)	57 (74.0)	51 (79.7)	0.857 (1) 0.486 (2)	18 (69.2)	38 (88.4)	39 (70.9)	0.060 (1) 0.757 (2)
	GA	5 (20.0)	10 (33.3)	10 (32.3)		33 (22.1)	19 (24.7)	13 (20.3)		8 (30.8)	4 (9.3)	15 (27.3)	
	GG	1 (4.0)	2 (6.7)	0 (0)		3 (2.0)	1 (1.3)	0 (0)		0 (0)	1 (2.3)	1 (1.8)	
	A	43 (86)	46 (76.7)	52 (83.9)	0.215 (1) 0.755 (2)	259 (86.9)	133 (86.4)	115 (89.8)	0.870 (1) 0.397 (2)	44 (84.6)	80 (93)	93 (84.6)	0.113 (1) 0.991 (2)
	G	7 (14)	14 (23.3)	10 (16.1)		39 (13.1)	21 (13.6)	13 (10.2)		8 (15.4)	6 (7)	17 (15.4)	
rs10168266	CC	14 (56.0)	17 (56.7)	18 (58.1)	0.956 (1) 0.972 (2)	101 (67.8)	53 (68.8)	38 (59.4)	0.528 (1) 0.234 (2)	18 (69.2)	30 (69.8)	42 (76.1)	0.934 (1) 0.777 (2)
	CT	9 (36.0)	10 (33.3)	11 (35.5)		42 (28.2)	23 (29.9)	20 (31.3)		7 (26.9)	12 (27.9)	11 (20.0)	
	TT	2 (8.0)	3 (10.0)	2 (6.5)		6 (4.0)	1 (1.3)	6 (9.4)		1 (3.8)	1 (2.3)	2 (3.6)	
	C	37 (74)	44 (73.3)	47 (75.8)	0.937 (1) 0.826 (2)	244 (81.9)	129 (83.8)	96 (75)	0.617 (1) 0.105 (2)	43 (82.7)	72 (83.7)	95 (86.4)	0.875 (1) 0.539 (2)
	T	13 (26)	16 (26.7)	15 (24.2)		54 (18.1)	25 (16.2)	32 (25)		9 (17.3)	14 (16.3)	15 (13.6)	

p—Significance level and Bonferroni-corrected significance level when $p < 0.05/4$; SNP—single nucleotide polymorphism; AMD—age-related macular degeneration. (1)—Early AMD vs. control group; (2)—exudative AMD vs. control group.

3.5. *STAT4* (rs10181656, rs7574865, rs7601754 and rs10168266) Genotype and Allele Associations with Early and Exudative AMD by Age

No statistically significant association was found between *STAT4* (rs10181656, rs7574865, rs7601754 and rs10168266) in early AMD, exudative AMD and control groups in ≤65-year-old subjects (Table 11).

Table 11. Binomial logistic regression analysis of *STAT4* (rs10181656, rs7574865, rs7601754, and rs10168266) in the early AMD, exudative AMD, and control groups in the ≤65-year-old subjects.

		≤65 y/o		
Model	**Genotype/Allele**	**OR (95% CI) ***	***p*-Value**	**AIC**
		Early AMD		
		rs10181656		
Codominant	CG vs. CC	1.040 (0.339–3.190)	0.945	79.720
	GG vs. CC	1.300 (0.187–9.021)	0.791	
Dominant	CG+GG vs. CC	1.083 (0.375–3.133)	0.883	77.769
Recessive	GG vs. CG+CC	1.278 (0.196–8.321)	0.798	77.724

Table 11. *Cont.*

		≤65 y/o		
Model	**Genotype/Allele**	**OR (95% CI) ***	***p*-Value**	**AIC**
Overdominant	CG vs. CC+GG	1.000 (0.338–2.955)	1.000	77.791
Additive	C	1.099 (0.486–2.486)	0.821	77.740
		rs7574865		
Codominant	GT vs. GG	0.841 (0.274–2.579)	0.762	79.664
	TT vs. GG	0.765 (0.095–6.175)	0.801	
Dominant	GT+TT vs. GG	0.828 (0.285–2.406)	0.729	77.671
Recessive	TT vs. GG+GT	0.821 (0.107–6.293)	0.850	77.755
Overdominant	GT vs. GG+TT	0.868 (0.292–2.587)	0.800	77.727
Additive	G	0.859 (0.369–1.999)	0.725	77.667
		rs7601754		
Codominant	GA vs. AA	2.111 (0.604–7.385)	0.242	78.180
	GG vs. AA	2.111 (0.176–25.349)	0.556	
Dominant	GA+GG vs. AA	2.111 (0.653–6.823)	0.212	76.180
Recessive	GG vs. AA+GA	1.714 (0.146–20.097)	0.668	77.598
Overdominant	GA vs. AA+GG	2.000 (0.579–6.908)	0.273	76.547
Additive	A	1.766 (0.673–4.634)	0.248	76.375
		rs10168266		
Codominant	CT vs. CC	0.915 (0.291–2.876)	0.879	79.701
	TT vs. CC	1.235 (0.180–8.459)	0.830	
Dominant	CT+TT vs. CC	0.973 (0.334–2.838)	0.960	77.789
Recessive	TT vs. CC+CT	1.278 (0.196–8.321)	0.798	77.724
Overdominant	CT vs. CC+TT	0.889 (0.291–2.711)	0.836	77.748
Additive	C	1.031 (0.459–2.317)	0.940	77.785
		Exudative AMD		
		rs10181656		
Codominant	CG vs. CC	0.765 (0.246–2.381)	0.643	80.418
	GG vs. CC	1.529 (0.242–9.674)	0.652	
Dominant	CG+GG vs. CC	0.892 (0.310–2.566)	0.832	78.944
Recessive	GG vs. CG+CC	1.704 (0.286–10.165)	0.559	78.633
Overdominant	CG vs. CC+GG	0.714 (0.238–2.143)	0.548	78.628
Additive	C	1.046 (0.480–2.279)	0.910	78.976
		rs7574865		
Codominant	GT vs. GG	0.765 (0.246–2.381)	0.643	80.418
	TT vs. GG	1.529 (0.242–9.674)	0.652	
Dominant	GT+TT vs. GG	0.892 (0.310–2.566)	0.832	78.944
Recessive	TT vs. GG+GT	1.704 (0.286–10.165)	0.559	78.633
Overdominant	GT vs. GG+TT	0.714 (0.238–2.143)	0.548	78.628
Additive	G	1.046 (0.480–2.279)	0.910	78.976

Table 11. *Cont.*

Model	Genotype/Allele	OR (95% CI) *	*p*-Value	AIC
		≤65 y/o		
		rs7601754		
Codominant	GA vs. AA	1.810 (0.524–6.253)	0.349	78.447
	GG vs. AA	-	-	
Dominant	GA+GG vs. AA	1.508 (0.460–4.943)	0.498	78.522
Recessive	GG vs. AA+GA	-	-	-
Overdominant	GA vs. AA+GG	1.905 (0.553–6.555)	0.307	77.909
Additive	A	1.190 (0.408–3.467)	0.750	78.886
		rs10168266		
Codominant	CT vs. CC	0.951 (0.309–2.926)	0.930	80.931
	TT vs. CC	0.778 (0.097–6.230)	0.813	
Dominant	CT+TT vs. CC	0.919 (0.317–2.664)	0.877	78.964
Recessive	TT vs. CC+CT	0.793 (0.104–6.069)	0.823	78.939
Overdominant	CT vs. CC+TT	0.978 (0.326–2.935)	0.968	78.987
Additive	C	0.912 (0.394–2.112)	0.830	78.942

p—Significance level and Bonferroni-corrected significance level when $p < 0.05/4$; OR—odds ratio; CI—confident interval; AIC—Akaike information criteria. * 95% confidence interval (CI) of the mean is a range with an upper and lower number calculated.

Furthermore, no statistically significant association was found between *STAT4* (rs10181656, rs7574865, rs7601754, and rs10168266) in the early AMD, exudative AMD, and control groups among subjects aged > 65 to ≤75 years (Table 12).

Table 12. Binomial logistic regression analysis of *STAT4* (rs10181656, rs7574865, rs7601754, and rs10168266) in the early AMD, exudative AMD, and control groups in the >65–≤75-year-old subjects.

Model	Genotype/Allele	OR (95% CI) *	*p*-Value	AIC
		>65 y/o–≤75 y/o		
		Early AMD		
		rs10181656		
Codominant	CG vs. CC	1.004 (0.555–1.816)	0.989	292.420
	GG vs. CC	0.464 (0.125–1.723)	0.251	
Dominant	CG+GG vs. CC	0.896 (0.509–1.577)	0.704	291.815
Recessive	GG vs. CG+CC	0.463 (0.127–1.692	0.244	290.420
Overdominant	CG vs. CC+GG	1.073 (0.598–1.924)	0.814	291.904
Additive	C	0.834 (0.529–1.316)	0.436	291.342
		rs7574865		
Codominant	GT vs. GG	0.966 (0.531–1.755)	0.909	293.285
	TT vs. GG	0.618 (0.189–2.021)	0.426	
Dominant	GT+TT vs. GG	0.896 (0.509–1.577)	0.704	291.815
Recessive	TT vs. GG+GT	0.626 (0.195–2.009)	0.431	291.298
Overdominant	GT vs. GG+TT	1.012 (0.562–1.821)	0.969	291.958
Additive	G	0.867 (0.554–1.357)	0.534	291.567

Table 12. *Cont.*

		>65 y/o–≤75 y/o		
Model	**Genotype/Allele**	**OR (95% CI) ***	***p*-Value**	**AIC**
		rs7601754		
Codominant	GA vs. AA	1.141 (0.597–2.182)	0.689	293.644
	GG vs. AA	0.661 (0.067–6.496)	0.722	
Dominant	GA+GG vs. AA	1.101 (0.585–2.073)	0.765	291.871
Recessive	GG vs. AA+GA	0.640 (0.065–6.261)	0.702	291.803
Overdominant	GA vs. AA+GG	1.152 (0.603–2.198)	0.669	291.778
Additive	A	1.049 (0.593–1.854)	0.871	291.933
		rs10168266		
Codominant	CT vs. CC	1.044 (0.568–1.916)	0.891	292.491
	TT vs. CC	0.318 (0.037–2.707)	0.294	
Dominant	CT+TT vs. CC	0.953 (0.527–1.723)	0.873	291.934
Recessive	TT vs. CC+CT	0.314 (0.037–2.653)	0.287	290.510
Overdominant	CT vs. CC+TT	1.085 (0.593–1.986)	0.791	291.890
Additive	C	0.876 (0.521–1.473)	0.617	291.707
		Exudative AMD		
		rs10181656		
Codominant	CG vs. CC	1.449 (0.771–2.720)	0.249	262.633
	GG vs. CC	1.622 (0.587–4.481)	0.351	
Dominant	CG+GG vs. CC	1.483 (0.823–2.674)	0.190	260.678
Recessive	GG vs. CG+CC	1.402 (0.525–3.743)	0.500	261.955
Overdominant	CG vs. CC+GG	1.349 (0.734–2.479)	0.335	261.476
Additive	C	1.333 (0.860–2.067)	0.199	260.763
		rs7574865		
Codominant	GT vs. GG	1.495 (0.794–2.816)	0.213	261.862
	TT vs. GG	1.914 (0.716–5.118)	0.196	
Dominant	GT+TT vs. GG	1.579 (0.876–2.847)	0.129	260.086
Recessive	TT vs. GG+GT	1.631 (0.633–4.205)	0.311	261.405
Overdominant	GT vs. GG+TT	1.349 (0.734–2.479)	0.335	261.476
Additive	G	1.419 (0.920–2.190)	0.113	259.911
		rs7601754		
Codominant	GA vs. AA	0.873 (0.424–1.797)	0.712	262.097
	GG vs. AA	-	-	
Dominant	GA+GG vs. AA	0.800 (0.391–1.636)	0.541	262.017
Recessive	GG vs. AA+GA	-	-	-
Overdominant	GA vs. AA+GG	0.896 (0.436–1.843)	0.765	262.308
Additive	A	0.749 (0.384–1.462)	0.397	261.651
		rs10168266		
Codominant	CT vs. CC	1.266 (0.661–2.425)	0.478	261.680
	TT vs. CC	2.658 (0.807–8.750)	0.108	
Dominant	CT+TT vs. CC	1.440 (0.786–2.638)	0.238	261.018
Recessive	TT vs. CC+CT	2.466 (0.764–7.960)	0.131	260.179
Overdominant	CT vs. CC+TT	1.158 (0.612–2.191)	0.652	262.196
Additive	C	1.454 (0.902–2.344)	0.124	260.067

p—Significance level and Bonferroni-corrected significance level when $p < 0.05/4$; OR—odds ratio; CI—confident interval; AIC—Akaike information criteria. * 95% confidence interval (CI) of the mean is a range with an upper and lower number calculated.

The binomial logistic regression analysis of *STAT4* (rs10181656, rs7574865, rs7601754, and rs10168266) in the early AMD, exudative AMD, and control groups in the >75-year-old subjects did not reveal any statistically significant association (Table 13).

Table 13. Binomial logistic regression analysis of *STAT4* (rs10181656, rs7574865, rs7601754, and rs10168266) in the early AMD, exudative AMD, and control groups in the >75-year-old subjects.

		>75 y/o		
Model	**Genotype/Allele**	**OR (95% CI) ***	***p*-Value**	**AIC**
		Early AMD		
		rs10181656		
Codominant	CG vs. CC	1.101 (0.364–3.331)	0.865	94.981
	GG vs. CC	0.593 (0.107–3.295)	0.550	
Dominant	CG+GG vs. CC	0.948 (0.348–2.586)	0.917	93.412
Recessive	GG vs. CG+CC	0.575 (0.107–3.087)	0.519	93.010
Overdominant	CG vs. CC+GG	1.176 (0.398–3.477)	0.769	93.336
Additive	C	0.873 (0.415–1.833)	0.719	93.294
		rs7574865		
Codominant	GT vs. GG	1.101 (0.364–3.331)	0.865	94.981
	TT vs. GG	0.593 (0.107–3.295)	0.550	
Dominant	GT+TT vs. GG	0.948 (0.348–2.586)	0.917	93.412
Recessive	TT vs. GG+GT	0.575 (0.107–3.087)	0.519	93.010
Overdominant	GT vs. GG+TT	1.176 (0.398–3.477)	0.769	93.336
Additive	G	0.873 (0.415–1.833)	0.719	93.294
		rs7601754		
Codominant	GA vs. AA	0.237 (0.063–0.891)	0.033	89.606
	GG vs. AA	-	-	
Dominant	GA+GG vs. AA	0.296 (0.085–1.034)	0.056	89.653
Recessive	GG vs. AA+GA	-	-	-
Overdominant	GA vs. AA+GG	0.231 (0.061–0.867)	0.030	88.373
Additive	A	0.422 (0.137–1.303)	0.134	91.082
		rs10168266		
Codominant	CT vs. CC	1.029 (0.342–3.091)	0.960	95.291
	TT vs. CC	0.600 (0.035–10.195)	0.724	
Dominant	CT+TT vs. CC	0.975 (0.339–2.806)	0.963	93.420
Recessive	TT vs. CC+CT	0.595 (0.036–9.943)	0.718	93.293
Overdominant	CT vs. CC+TT	1.051 (0.352–3.135)	0.929	93.415
Additive	C	0.930 (0.372–2.321)	0.876	93.398
		Exudative AMD		
		rs10181656		
Codominant	CG vs. CC	1.548 (0.544–4.410)	0.413	104.092
	GG vs. CC	0.516 (0.093–2.854)	0.448	

Table 13. *Cont.*

		>75 y/o		
Model	**Genotype/Allele**	**OR (95% CI) ***	***p*-Value**	**AIC**
Dominant	CG+GG vs. CC	1.239 (0.478–3.213)	0.660	103.478
Recessive	GG vs. CG+CC	0.442 (0.083–2.359)	0.339	102.779
Overdominant	CG vs. CC+GG	1.676 (0.603–4.664)	0.322	102.661
Additive	C	0.977 (0.467–2.044)	0.952	103.669
		rs7574865		
Codominant	GT vs. GG	1.500 (0.528–4.265)	0.447	103.326
	TT vs. GG	0.333 (0.051–2.200)	0.254	
Dominant	GT+TT vs. GG	1.150 (0.443–2.987)	0.774	103.590
Recessive	TT vs. GG+GT	0.289 (0.045–1.849)	0.190	101.918
Overdominant	GT vs. GG+TT	1.676 (0.603–4.664)	0.322	102.661
Additive	G	0.886 (0.416–1.888)	0.754	103.576
		rs7601754		
Codominant	GA vs. AA	0.865 (0.311–2.409)	0.782	104.817
	GG vs. AA	-	-	
Dominant	GA+GG vs. AA	0.923 (0.334–2.550)	0.877	103.649
Recessive	GG vs. AA+GA	-	-	-
Overdominant	GA vs. AA+GG	0.844 (0.303–2.346)	0.745	103.568
Additive	A	1.006 (0.386–2.618)	0.990	103.673
		rs10168266		
Codominant	CT vs. CC	0.673 (0.225–2.017)	0.480	105.180
	TT vs. CC	0.857 (0.073–10.064)	0.902	
Dominant	CT+TT vs. CC	0.696 (0.246–1.969)	0.495	103.214
Recessive	TT vs. CC+CT	0.943 (0.082–10.901)	0.963	103.671
Overdominant	CT vs. CC+TT	0.679 (0.228–2.018)	0.486	103.195
Additive	C	0.778 (0.331–1.824)	0.563	103.344

p—Significance level and Bonferroni-corrected significance level when $p < 0.05/4$; OR—odds ratio; CI—confident interval; AIC—Akaike information criteria. * 95% confidence interval (CI) of the mean is a range with an upper and lower number calculated.

Serum STAT4 levels were measured in patients with exudative AMD (n = 40) and in the control group (n = 40). We found that exudative AMD patients had lower STAT4 serum levels when compared to the control group (median (IQR): 0.118 (0.042) vs. 0.262 (0.385), $p = 0.005$). The results are shown in Figure 1.

However, no statistically significant differences were observed in the analysis of STAT4 levels between the early AMD and control groups (mean (std. deviation): 0.164 (0.068) vs. 0.859 (2.122), $p = 0.226$) (Figure 2).

A comparison of the serum STAT4 levels among different genotypes for selected single nucleotide polymorphisms was performed. The exudative AMD patients with at least one G allele of the *STAT4* rs10181656 had lower serum STAT4 levels than the control group subjects ($p = 0.011$). Also, the exudative AMD patients with at least one T allele of *STAT4* rs10168266 had lower serum STAT4 levels than the control group subjects ($p = 0.039$) (Table 14).

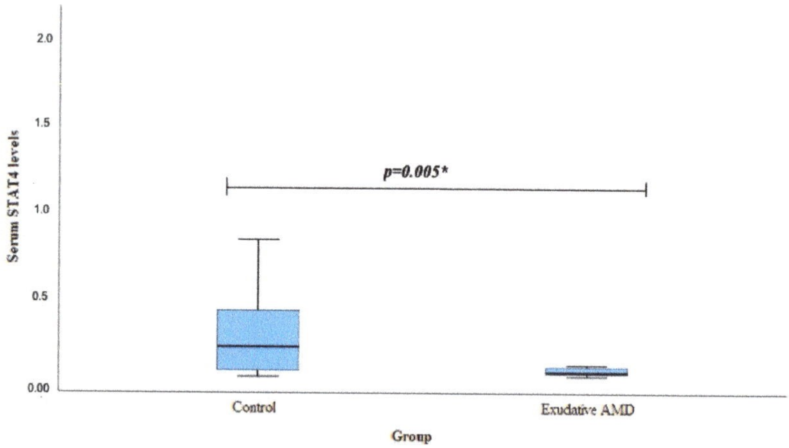

Figure 1. Serum levels of STAT4 in the control group and patients with exudative AMD. AMD: age-related macular degeneration; STAT4: signal transducer and activator of transcription 4. * The Mann–Whitney U test was used.

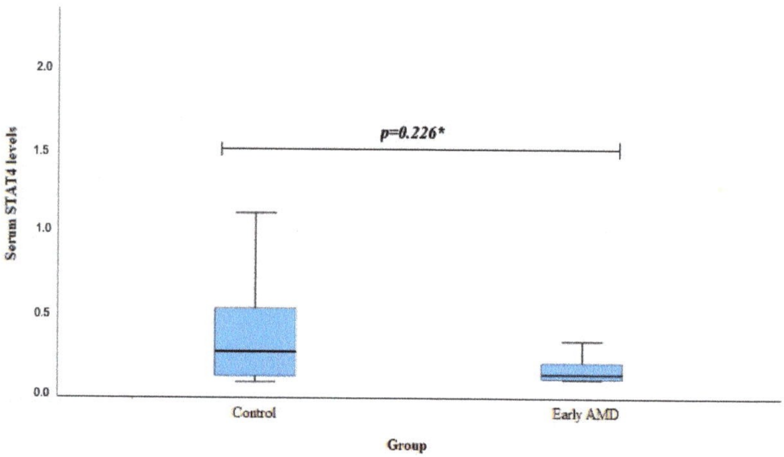

Figure 2. Serum levels of STAT4 in the control group and in patients with early AMD. AMD: age-related macular degeneration; STAT4: signal transducer and activator of transcription 4. * Student's *t*-test was used.

Table 14. Serum STAT4 level associations with STAT4 SNPs.

Genotype	Serum STAT4 Levels			*p*-Value
	Early AMD Mean (Std. Deviation)	Exudative AMD Median (IQR)	Control Median (IQR)	
		STAT4 rs10181656		
CC	0.431 (-)	0.109 (0.038)	0.202 (0.560)	0.555 [1] 0.405 [2]
CG+GG	0.256 (0.198)	0.178 (0.823)	0.292 (0.209)	0.826 [1] 0.011 [2]

Table 14. *Cont.*

Genotype	Serum STAT4 Levels			p-Value
	Early AMD Mean (Std. Deviation)	Exudative AMD Median (IQR)	Control Median (IQR)	
		STAT4 rs7574865		
GG	0.182 (0.198)	0.194 (0.123)	0.324 (0.318)	0.728 [1] 0.054 [2]
GT+TT	0.356 (0.188)	0.198 (0.353)	0.292 (0.209)	0.756 [1] 0.062 [2]
		STAT4 rs7601754		
AA	0.305 (0.187)	0.221 (0.198)	0.458 (0.268)	0.631 [1] 0.858 [2]
GA+GG	0.176 (0.124)	0.174 (0.155)	0.258 (0.144)	0.972 [1] 0.889 [2]
		STAT4 rs10168266		
CC	0.256 (0.266)	0.152 (0.190)	0.268 (0.268)	0.821 [1] 0.658 [2]
CT+TT	0.255 (0.178)	0.185 (0.562)	0.268 (0.154)	0.956 [1] 0.039 [2]

[1] Early AMD vs. control group. [2] Exudative AMD vs. control group.

4. Discussion

We performed a study investigating the associations of the single nucleotide polymorphisms rs10181656, rs7574865, rs7601754, and rs10168266 of the *STAT4* gene with AMD. A total of 500 subjects participated in the study: 150 with early AMD, 150 with exudative AMD, and 200 healthy subjects. As far as we are aware, no scientific studies have been conducted to investigate the impact of these polymorphisms on AMD. The scientific literature states that STAT4 is involved in the pathogenesis of various autoimmune and inflammatory diseases, and the *STAT4* polymorphisms rs10181656, rs7574865, rs7601754, and rs10168266 are associated with various autoimmune diseases, such as rheumatoid arthritis (RA), systemic lupus erythematosus (SLE), and systemic sclerosis (SS) [20].

The analysis of the genotype and allele distribution of the single nucleotide polymorphisms rs10181656, rs7574865, rs7601754, and rs10168266 of *STAT4* did not reveal statistically significant data in our research. There are few studies on the *STAT4* rs10181656 SNP in the scientific literature databases. Scientific databases state that this SNP is associated with RA, SS, and SLE [20–22]. In a 2017–2018 study conducted in Iran, the *STAT4* rs10181656 polymorphism was strongly associated with RA risk ($p = 0.007$), but no association was found with SS [21]. The *STAT4* rs10181656 polymorphism has been associated with autoimmune diseases such as rheumatoid arthritis, systemic sclerosis, and systemic lupus erythematosus [21,22]. Studies investigating the association of the *STAT4* rs10181656 SNP with the above diseases can be found in the literature. One such study was conducted in Mexico and included 869 Mexican subjects, including 415 with RA, 128 with SLE, and 326 healthy controls. After investigating the association between *STAT4* rs7574865G/T polymorphism and the mentioned diseases, the *STAT4* rs7574865 G/T genotype was found to be associated with both RA and SLE risk [23].

In 2016, Greek researchers published a study that found that the GG genotype and the G allele of *STAT4* rs10181656 were significantly associated with the occurrence of psoriatic arthritis [24]. Ni Yan and coauthors investigated the association of *STAT4* polymorphisms with autoimmune thyroid diseases. The study, published in 2014, concluded that the *STAT4* rs7574685 T allele and the *STAT4* rs10181656 G allele were statistically significantly associated with the occurrence of thyroid autoimmune diseases such as Graves and Hashimototis in the Chinese Han population [25]. Most studies conducted on the *STAT4* rs7574865 SNP investigated the association of this polymorphism with RA and SLE. EnPeng GU and coauthors conducted a study systematizing data from 28 case–control studies examining the association between the *STAT4* rs7574865 polymorphism and RA. The results showed that the *STAT4* rs7574865 TT genotype, GT+TT genotype, and T allele were significantly associated with RA in European, Asian, South American, and African groups [26]. Junfeng Zheng and co-authors conducted a systematic review study, in 2013, which showed that the *STAT4* rs7574865 SNP is associated with three autoimmune diseases, RA, SS, and SLE, and that the *STAT4* rs7574865 T allele increases the probability of disease [20]. According to the results of a study by Ya-ling Liang and co-authors, the *STAT4* rs7574865 SNP is not only associated with RA, SS, and SLE but also slightly associated with the risk of type 1 diabetes, juvenile idiopathic arthritis (JIA), and ulcerative colitis (UC) [27]. A study by Hui Yuan and co-authors reports similar findings to the studies previously discussed: the *STAT4* rs7574865 polymorphism is statistically significantly associated with SLE in European and Asian groups. Still, no statistically significant data show that the *STAT4* rs7601754 SNP is associated with SLE, although the authors do not rule out this possibility [28]. The *STAT4* rs10168266 SNP is also associated with SLE risk. The association of the *STAT4* rs10168266 SNP with SLE is confirmed by a study conducted by Malaysian researchers, which reported that the *STAT4* rs10168266 SNP is significantly associated with the development of SLE in the Malaysian population [29]. A total of 790 Malaysian citizens participated in the study, of which 360 were SLE patients and 430 were healthy controls.

The *STAT4* polymorphism rs7574865 has been associated with autoimmune and liver diseases, such as hepatocellular carcinoma (HCC), chronic hepatitis B (CHB), and liver cirrhosis (LC) [30,31]. In 2022, a study was published in the PubMed database that was conducted among a Chinese Han population of 3151 subjects, of whom 968 had chronic hepatitis B, 316 had liver cirrhosis, and 1021 had hepatocellular carcinoma. The control group consisted of 846 healthy subjects. The research results suggest that the GG genotype of the *STAT4* polymorphism rs7574865 is significantly associated with the risk of HCC, LC, and CHB [30]. Gao Wenyan and co-authors published a study, in 2019, that aimed to systematically examine the association of the *STAT4* rs7574865 SNP with RA, including across ethnic groups. The systematic data showed that the T allele of the *STAT4* rs7574865 polymorphism is associated with the risk of RA in European and Asian populations but in groups of individuals aged 50–60 years on average [32]. The previously mentioned SNP (i.e., rs7574865) is also associated with systemic lupus erythematosus. The analysis by Jia-Min Wang and co-authors aimed to determine the associations of *STAT4* polymorphisms rs10168266 and rs7574865 with SLE risk. In the aforementioned study, the genotypes TT and CT of the *STAT4* polymorphisms rs10168266 and rs7574865 were found to be associated with SLE risk [33]. In the literature, rs7574865 is also associated with type 1 diabetes. The T allele and GT genotype of the *STAT4* rs7574865 polymorphism were statistically significantly associated with type 1 diabetes in Egyptian study patients [34]. In the scientific literature, there are few studies on the *STAT4* polymorphism rs7601754. One study in the PubMed database examined the associations of the *STAT4* rs7574865 and the *STAT4* rs7601754 SNP with SLE. The results presented stated that the mentioned *STAT4* rs7574865 polymorphism was associated with an increased risk of SLE, whereas the *STAT4* rs7601754 polymorphism was associated with a reduced risk of SLE [35].

There are also very few studies on the *STAT4* rs10168266 SNP. A literature review revealed that this polymorphism has been studied in the scientific literature databases mainly in the search for associations with SLE. A study conducted by Japanese researchers on 308 SLE patients and 306 control patients concluded that the *STAT4* polymorphisms rs7574865, rs11889341, and the aforementioned rs10168266 were associated with SLE risk. It was also found that the association between the above polymorphism and SLE was higher in the Japanese population than in the European or American populations [36]. Studies conducted by researchers associate single nucleotide polymorphisms of the *STAT4* gene rs10181656, rs7574865, rs7601754, and rs10168266 with various autoimmune diseases.

In our study, we investigated the association between the *STAT4* rs10181656, rs7574865, rs7601754, and rs10168266 single nucleotide polymorphisms (SNPs) and AMD. Contrary to our expectations, no significant associations were found, but it is important to note that these specific *STAT4* gene variants have not previously been studied, making definitive conclusions challenging.

Interestingly, our analysis of serum STAT4 levels revealed noteworthy findings. Exudative AMD patients carrying at least one G allele of *STAT4* rs10181656 exhibited significantly lower serum STAT4 levels than subjects in the control group ($p = 0.011$). Similarly, those with at least one T allele of *STAT4* rs10168266 had lower serum STAT4 levels than the control group subjects ($p = 0.039$). This observation suggests a potential link between specific STAT4 genotypes and decreased serum STAT4 levels in individuals with exudative AMD. Moreover, our study demonstrated a broader significance by revealing significantly lower overall serum STAT4 levels in exudative AMD patients compared to the control group ($p = 0.005$). This implies a potential association between decreased STAT4 levels and the presence of exudative AMD. However, no significant differences were observed in the STAT4 levels between the early AMD and control groups ($p = 0.226$). This suggests a nuanced role of STAT4, specifically implicated in the exudative stage of AMD, indicating a distinct association of STAT4 with different phases of AMD.

These findings not only contribute to our understanding of the genetic and serum level factors associated with AMD but also highlight the need for further exploration of the role of STAT4 in the context of different AMD stages.

While shedding light on the potential link between *STAT4* gene polymorphisms and AMD, our study has notable limitations. The existing literature lacks a definitive consensus on the role of *STAT4* in AMD pathogenesis, introducing an element of uncertainty. We acknowledge the omission of environmental factors, such as chronic light damage and aging, in our focus. Additionally, the use of serum STAT4 levels as a proxy for retinal tissue expression presents a limitation, prompting consideration for direct assessments in future research. The modest sample size, though offering initial insight, underscores the need for larger cohorts to validate our findings. Despite these limitations, our study contributes to the ongoing discourse on AMD, and we are committed to refining our approach in future investigations for a more comprehensive understanding.

Author Contributions: Conceptualization, G.G. and R.L.; Methodology, G.G.; Formal analysis, T.B., G.G. and K.K.; Investigation, T.B.; Resources, L.K. and D.Z.; Data curation, G.G. and K.K.; Writing–original draft, T.B., G.G., K.K. and R.L.; Writing–review & editing, G.G., K.K. and R.L.; Visualization, G.G.; Supervision, R.L.; Funding acquisition, L.K. All authors have read and agreed to the published version of the manuscript.

Funding: This research received no external funding.

Institutional Review Board Statement: The study was conducted in accordance with the Declaration of Helsinki, and approved by The Kaunas Regional Biomedical Research Ethics Committee approved the study (approval numbers: 9 July 2015 No. BE-2-26 and 26 January 2017 No. P1-BE-2-26/2015).

Informed Consent Statement: Informed consent was obtained from all subjects involved in the study.

Data Availability Statement: Data will be sent upon a request.

Conflicts of Interest: The authors declare no conflict of interest.

References

1. García-Layana, A.; Cabrera-López, F.; García-Arumí, J.; Arias-Barquet, L.; Ruiz-Moreno, J.M. Early and intermediate age-related macular degeneration: Update and clinical review. *Clin. Interv. Aging* **2017**, *12*, 1579. [CrossRef] [PubMed]
2. Cho, Y.K.; Park, D.H.; Jeon, I.C. Medication Trends for Age-Related Macular Degeneration. *Int. J. Mol. Sci.* **2021**, *22*, 11837. [CrossRef] [PubMed]

3. Schnabolk, G. Systemic Inflammatory Disease and AMD Comorbidity. In *Advances in Experimental Medicine and Biology*; Springer: Cham, Switzerland, 2019; Volume 1185, pp. 27–31. [CrossRef]
4. Fan, Q.; Maranville, J.C.; Fritsche, L.; Sim, X.; Cheung, C.M.G.; Chen, L.J.; Gorski, M.; Yamashiro, K.; Ahn, J.; Laude, A.; et al. HDL-cholesterol levels and risk of age-related macular degeneration: A multiethnic genetic study using Mendelian randomization. *Int. J. Epidemiol.* **2017**, *46*, 1891–1902. [CrossRef] [PubMed]
5. Zhang, Q.Y.; Tie, L.J.; Wu, S.S.; Lv, P.L.; Huang, H.W.; Wang, W.Q.; Wang, H.; Ma, L. Overweight, Obesity, and Risk of Age-Related Macular Degeneration. *Investig. Ophthalmol. Vis. Sci.* **2016**, *57*, 1276–1283. [CrossRef] [PubMed]
6. Deng, Y.; Qiao, L.; Du, M.; Qu, C.; Wan, L.; Li, J.; Huang, L. Age-related macular degeneration: Epidemiology, genetics, pathophysiology, diagnosis, and targeted therapy. *Genes Dis.* **2021**, *9*, 62–79. [CrossRef] [PubMed]
7. Bruneau, S.; Nakayama, H.; Woda, C.B.; Flynn, E.A.; Briscoe, D.M. DEPTOR regulates vascular endothelial cell activation and proinflammatory and angiogenic responses. *Blood* **2013**, *122*, 1833–1842. [CrossRef]
8. Qing, M.; Schumacher, K.; Heise, R.; Wöltje, M.; Vazquez-Jimenez, J.F.; Richter, T.; Arranda-Carrero, M.; Hess, J.; von Bernuth, G.; Seghaye, M.-C. Intramyocardial synthesis of pro- and anti-inflammatory cytokines in infants with congenital cardiac defects. *J. Am. Coll. Cardiol.* **2003**, *41*, 2266–2274. [CrossRef]
9. Torpey, N.; Maher, S.E.; Bothwell, A.L.; Pober, J.S. Interferon alpha but not interleukin 12 activates STAT4 signalling in human vascular endothelial cells. *J. Biol. Chem.* **2004**, *279*, 26789–26796. [CrossRef]
10. Liang, Y.; Pan, H.-F.; Ye, D.-Q. Therapeutic potential of STAT4 in autoimmunity. *Expert Opin. Ther. Targets* **2014**, *18*, 945–960. [CrossRef]
11. Darnell, J.E., Jr. *Signal Transducers and Activators of Transcription (STATs)*; Springer: Dordrecht, The Netherlands, 2003.
12. Loh, C.Y.; Arya, A.; Naema, A.F.; Wong, W.F.; Sethi, G.; Looi, C.Y. Signal transducer and activator of transcription (STATs) proteins in cancer and inflammation: Functions and therapeutic implication. *Front. Oncol.* **2019**, *9*, 48. [CrossRef]
13. Vidangos, N.; Maris, A.E.; Young, A.; Hong, E.; Pelton, J.G.; Batchelor, J.D.; Wemmer, D.E. Structure, function, and tethering of DNA-binding domains in σ54 transcriptional activators. *Biopolymers* **2013**, *99*, 1082. [CrossRef] [PubMed]
14. Yang, C.; Mai, H.; Peng, J.; Zhou, B.; Hou, J.; Jiang, D. STAT4: An immunoregulator contributing to diverse human diseases. *Int. J. Biol. Sci.* **2020**, *16*, 1575–1585. [CrossRef] [PubMed]
15. Mehrpouya-Bahrami, P.; Moriarty, A.K.; de Melo, P.; Keeter, W.C.; Alakhras, N.S.; Nelson, A.S.; Hoover, M.; Barrios, M.S.; Nadler, J.L.; Serezani, C.H.; et al. STAT4 is expressed in neutrophils and promotes antimicrobial immunity. *JCI Insight* **2021**, *6*, e141326. [CrossRef] [PubMed]
16. Cui, J.; Tong, R.; Xu, J.; Tian, Y.; Pan, J.; Wang, N.; Chen, H.; Peng, Y.; Fei, S.; Ling, W.; et al. Association between STAT4 gene polymorphism and type 2 diabetes risk in Chinese Han population. *BMC Med. Genom.* **2021**, *14*, 169. [CrossRef] [PubMed]
17. Cui, J.; Tong, R.; Xu, J.; Tian, Y.; Pan, J.; Wang, N.; Chen, H.; Peng, Y.; Fei, S.; Ling, W.; et al. STAT4 sequence variant and elevated gene expression are associated with type 1 diabetes in Polish children. *Cent. Eur. J. Immunol.* **2020**, *45*, 22–28. [CrossRef]
18. Shi, Z.; Zhang, Q.; Chen, H.; Lian, Z.; Liu, J.; Feng, H.; Miao, X.; Du, Q.; Zhou, H. *STAT4* Polymorphisms are Associated with Neuromyelitis Optica Spectrum Disorders. *Neuromol. Med.* **2017**, *19*, 493–500. [CrossRef] [PubMed]
19. Al-Zamil, W.M.; Yassin, S.A. Recent developments in age-related macular degeneration: A review. *Clin. Interv. Aging* **2017**, *12*, 1313–1330. [CrossRef] [PubMed]
20. Zheng, J.; Yin, J.; Huang, R.; Petersen, F.; Yu, X. Meta-analysis reveals an association of *STAT4* polymorphisms with systemic autoimmune disorders and anti-dsDNA antibody. *Hum. Immunol.* **2013**, *74*, 986–992. [CrossRef]
21. Ghanavati, F.; Nezhad, S.R.K.; Hajjari, M.R.; Akhoond, M.R. Association of Signal Transducer and Activator of Transcription 4 rs10181656 Polymorphism with Rheumatoid Arthritis and Systemic Sclerosis in Khuzestan Province in Southwestern Iran. *Arch. Rheumatol.* **2019**, *34*, 434–442. [CrossRef]
22. Hellquist, A.; Sandling, J.K.; Zucchelli, M.; Koskenmies, S.; Julkunen, H.; D'Amato, M.; Garnier, S.; Syvänen, A.-C.; Kere, J. Variation in STAT4 is associated with systemic lupus erythematosus in a Finnish family cohort. *Ann. Rheum. Dis.* **2010**, *69*, 883–886. [CrossRef]
23. Beltrán Ramírez, O.; Mendoza Rincón, J.F.; Barbosa Cobos, R.E.; Alemán Ávila, I.; Ramírez Bello, J. STAT4 confers risk for rheumatoid arthritis and systemic lupus erythematosus in Mexican patients. *Immunol. Lett.* **2016**, *175*, 40–43. [CrossRef] [PubMed]
24. Myrthianou, E.; Zervou, M.; Budu-Aggrey, A.; Eliopoulos, E.; Kardassis, D.; Boumpas, D.; Kougkas, N.; Barton, A.; Sidiropoulos, P.; Goulielmos, G. Investigation of the genetic overlap between rheumatoid arthritis and psoriatic arthritis in a Greek population. *Scand. J. Rheumatol.* **2016**, *46*, 180–186. [CrossRef]
25. Yan, N.; Meng, S.; Zhou, J.; Xu, J.; Muhali, F.S.; Jiang, W.; Shi, L.; Shi, X.; Zhang, J. Association between *STAT4* Gene Polymorphisms and Autoimmune Thyroid Diseases in a Chinese Population. *Int. J. Mol. Sci.* **2014**, *15*, 12280–12293. [CrossRef] [PubMed]
26. Gu, E.; Lu, J.; Xing, D.; Chen, X.; Xie, H.; Liang, J.; Li, L. Rs7574865 polymorphism in signal transducers and activators of transcription 4 gene and rheumatoid arthritis: An updated meta-analysis of 28 case-control comparisons. *Int. J. Rheum. Dis.* **2015**, *18*, 3–16. [CrossRef]
27. Liang, Y.L.; Wu, H.; Shen, X.; Li, P.Q.; Yang, X.Q.; Liang, L.; Tian, W.H.; Zhang, L.F.; Xie, X.D. Association of STAT4 rs7574865 polymorphism with autoimmune diseases: A meta-analysis. *Mol. Biol. Rep.* **2012**, *39*, 8873–8882. [CrossRef] [PubMed]
28. Yuan, H.; Feng, J.-B.; Pan, H.-F.; Qiu, L.-X.; Li, L.-H.; Zhang, N.; Ye, D.-Q. A meta-analysis of the association of STAT4 polymorphism with systemic lupus erythematosus. *Mod. Rheumatol.* **2014**, *20*, 257–262. [CrossRef]

29. Chai, H.C.; Chua, K.H.; Lim, S.K.; Phipps, M.E. Insight into gene polymorphisms involved in toll-like receptor/interferon signalling pathways for systemic lupus erythematosus in South East Asia. *J. Immunol. Res.* **2014**, *2014*, 529167. [CrossRef] [PubMed]
30. Yang, C.; Chen, H.; Zhou, B.; Yin, J.; Cao, G.; Hou, J.; Jiang, D. The effects of the interactions of *STAT4* rs7574865 with HBV mutations on the risk of hepatocellular carcinoma. *Mol. Carcinog.* **2022**, *61*, 933–940. [CrossRef]
31. Shi, H.; He, H.; Ojha, S.C.; Sun, C.; Fu, J.; Yan, M.; Deng, C.; Sheng, Y.J. Association of STAT3 and STAT4 polymorphisms with susceptibility to chronic hepatitis B virus infection and risk of hepatocellular carcinoma: A meta-analysis. *Biosci. Rep.* **2019**, *39*, 20190783. [CrossRef]
32. Gao, W.; Dong, X.; Yang, Z.; Mao, G.; Xing, W. Association between rs7574865 polymorphism in *STAT4* gene and rheumatoid arthritis: An updated meta-analysis. *Eur. J. Intern. Med.* **2020**, *71*, 101–103. [CrossRef]
33. Wang, J.M.; Xu, W.D.; Huang, A.F. Association of STAT4 Gene Rs7574865, Rs10168266 Polymorphisms and Systemic Lupus Erythematosus Susceptibility: A Meta-analysis. *Immunol. Investig.* **2020**, *50*, 282–294. [CrossRef] [PubMed]
34. Abdelmajed, S.S.; El-Dessouky, M.A.; SalahElDin, D.S.; Hassan, N.A.M.; Zaki, M.E.; Ismail, S. Assessing the association of rs7574865 STAT4 gene variant and type 1 diabetes mellitus among Egyptian patients. *J. Genet. Eng. Biotechnol.* **2021**, *19*, 112. [CrossRef] [PubMed]
35. Ebrahimiyan, H.; Rezaei, R.; Mostafaei, S.; Aslani, S.; Goulielmos, G.N.; Jamshidi, A.; Mahmoudi, M. Association study between STAT4 polymorphisms and susceptibility to systemic lupus erythematosus disease: A systematic review and meta-analysis. *Meta Gene* **2018**, *16*, 241–247. [CrossRef]
36. Kawasaki, A.; Ito, I.; Hikami, K.; Ohashi, J.; Hayashi, T.; Goto, D.; Matsumoto, I.; Ito, S.; Tsutsumi, A.; Koga, M.; et al. Role of *STAT4* polymorphisms in systemic lupus erythematosus in a Japanese population: A case-control association study of the STAT1-STAT4 region. *Arthritis Res. Ther.* **2008**, *10*, R113. [CrossRef]

biomedicines

MDPI

Article

Pharmacogenetic Association between Allelic Variants of the Autophagy-Related Genes and Anti-Vascular Endothelial Growth Factor Treatment Response in Neovascular Age-Related Macular Degeneration

Oyuna S. Kozhevnikova [1,*], Anzhella Zh. Fursova [1,2,3], Anna S. Derbeneva [1,2,3], Ida F. Nikulich [1,2,3], Vasiliy A. Devyatkin [1] and Nataliya G. Kolosova [1]

1 Federal Research Center Institute of Cytology and Genetics SB RAS, Pr. Lavrentiev, 10, 630090 Novosibirsk, Russia; fursova@bionet.nsc.ru (A.Z.F.); derbeneva@bionet.nsc.ru (A.S.D.); nikulich@bionet.nsc.ru (I.F.N.); devyatkin@bionet.nsc.ru (V.A.D.); kolosova@bionet.nsc.ru (N.G.K.)
2 State Novosibirsk Regional Clinical Hospital, St. Nemirovich-Danchenko, 130, 630087 Novosibirsk, Russia
3 Department of Ophthalmology, Novosibirsk State Medical University, Pr. Krasny, 52, 630091 Novosibirsk, Russia
* Correspondence: oidopova@bionet.nsc.ru

Abstract: Background: Age-related macular degeneration (AMD) is the leading cause of late-onset blindness in elderly. The occurrence and development of AMD is a multifactorial complex process where autophagy plays an important role. The first-line drugs for neovascular AMD (nAMD) are inhibitors of VEGF, with up to 30% of patients having an incomplete response to treatment. Genetic factors may influence the response to anti-VEGF therapy and explain treatment outcome variability. We aimed to estimate the role of polymorphic markers of the *MTOR* (rs1064261, rs1057079, rs11121704, rs2295080), *SQSTM1* (rs10277), *ULK1* (rs11246867, rs3088051), *MAP1LC3A* (rs73105013) and *ATG5* (rs573775) genes in the development of nAMD and the efficacy of anti-VEGF therapy response. Methods: Genotyping by allele-specific PCR was performed in 317 controls and 315 nAMD patients in the Russian population. Of them, 196 treatment-naive nAMD patients underwent three monthly intravitreal injections (IVIs) of aflibercept. Genotypic frequencies were compared with OCT markers of therapy effectiveness and best-corrected visual acuity (BCVA) measures. The main outcomes were the BCVA gain and decrease in central retinal thickness (CRT). Results: *MTOR*-rs1057079-C, *MTOR*-rs11121704-C and *MTOR*-rs2295080-G alleles were associated with an increased risk of nAMD. The BCVA was increased in 117 (59.7%) patients by 10 [5–20] letters, did not changed in 59 (30.1%), and was decreased in 20 (10.2%) patients. ULK1-rs3088051 was associated with BCVA change. Among patients with the TT and CT genotypes for *ULK1*-rs3088051, an improvement in visual acuity was noted in 67.6% and 53.8% of cases, while in patients with the CC genotype, an increase in BCVA was recorded in 37.5% of cases ($p = 0.01$). The decrease in CRT was associated with *SQSTM1*-rs10277 ($p = 0.001$): it was significantly higher in TT (93 [58–122] mkm) and CT (66 [30–105] mkm) carriers compared to the CC genotype (47 [24–68] mkm). Other SNPs did not show significant associations with the outcome of anti-VEGF treatment. Conclusions: *MTOR* gene polymorphisms are moderately associated with the risk of nAMD. *SQSTM1*-rs10277 and *ULK1*-rs3088051 may influence short-term response to intravitreal anti-VEGF treatment. The results suggest that autophagy could be a target for future drugs to overcome resistance to anti-VEGF therapy.

Keywords: age-related macular degeneration; autophagy; response to anti-VEGF therapy; pharmacogenetics; aflibercept

Citation: Kozhevnikova, O.S.; Fursova, A.Z.; Derbeneva, A.S.; Nikulich, I.F.; Devyatkin, V.A.; Kolosova, N.G. Pharmacogenetic Association between Allelic Variants of the Autophagy-Related Genes and Anti-Vascular Endothelial Growth Factor Treatment Response in Neovascular Age-Related Macular Degeneration. *Biomedicines* **2023**, *11*, 3079. https://doi.org/10.3390/biomedicines11113079

Academic Editor: Victor Chong

Received: 23 October 2023
Revised: 13 November 2023
Accepted: 15 November 2023
Published: 16 November 2023

1. Introduction

Age-related macular degeneration (AMD) is the leading cause of irreversible late-onset blindness in developed countries. A recent meta-analysis of epidemiologic studies estimated that around 25.3% of people aged over 60 years present early or intermediate AMD, while 2.4% present late AMD [1]. Due to population ageing, a number of AMD patients is expected to increase by a third in the coming decades [2]. AMD is a progressive disease affecting the macular area due to developing pathological processes in the retinal pigment epithelium (RPE), Bruch's membrane and choriocapillaries [3]. Because of the significant multifactorial nature and complexity of the disease, the exact mechanism of AMD pathogenesis remains uncertain. Most commonly, AMD starts in its dry form, which may progress to a wet form or neovascular AMD (nAMD) in 10–20% of cases [2]. The main symptom of nAMD is the macular neovascularization (MNV)—the pathological growth of new blood vessels into the different layers of the central retina, causing the accumulation of subretinal fluid (SRF), intraretinal fluid (IRF) and RPE detachment (PED) [4]. Without treatment, nAMD causes 90% of cases of severe vision loss [2]. Therefore, it is highly desirable to develop effective pharmacological treatment of macular degeneration [5].

The current standard of nAMD treatment is the antiangiogenic therapy targeting vascular endothelial growth factor (VEGF), which has been a successful breakthrough in the treatment of MNV. However, it requires repeated and costly intravitreal injections that cannot be postponed, as well as frequent ophthalmological examinations, making treatment expensive and time-consuming [6]. Moreover, despite its effectiveness, in some cases, there is a discrepancy between the expected clinical morphological and functional data and the obtained results: the therapeutic effect may be reduced or even absent. Up to 30% of patients show an unsatisfactory response to anti-VEGF treatment, suggesting additional factors are at work [7]. Variability in therapeutic response may be due to hereditary factors such as genetic polymorphism. Indeed, the studies indicate that in patients with nAMD, the response to anti-VEGF therapy depends on the genotype of genes in the complement system [6,8–12] and VEGF-related pathway [6,13–15]. Recently, Paterno et al. [16] found that single nucleotide polymorphisms (SNPs) of autophagy genes have been associated with nAMD and the outcomes of anti-VEGF treatment in a cohort of Finnish patients. The accumulated data indicate that genetic predisposition contributes to resistance to anti-VEGF therapy. However, genetic factors could be population-specific. Thus, studies that focused on the identification or replication of susceptibility genes in AMD development and on the response to treatments in different populations do not lose their relevance [12].

Autophagy is a lysosomal-dependent degradation process that is highly conserved and maintains cellular homeostasis by sequestering cytosolic material for degradation [17]. Dysfunctional autophagy leads to pathological accumulation of the cargo, which has been linked to a range of human diseases, including neurodegenerative diseases, infectious and autoimmune diseases, and various forms of cancer [18]. Recent studies proved the critical role of autophagy in the homeostasis of aging RPE cells [19]. Disturbances in waste clearance result in the accumulation of harmful lipid and protein aggregates, which can act as a physical barrier to intracellular transport and disrupt RPE cell function [20]. According to a number of characteristics, AMD can be attributed to 'autophagopathies'—a class of complex human diseases whose etiology is failure in the work of the autophagy machinery, whether directly or indirectly related to an abnormal flux in autophagy, LC3-associated phagocytosis or any associated trafficking [21]. Recently, autophagy has been implicated to cause tumor resistance to antiangiogenic therapy [17], which suggests an analogous connection between autophagy and anti-VEGF intravitreal injections during AMD treatment. Considering all of the above, we assumed that genetic variability in autophagy pathway genes may influence the risk of nAMD and the response to anti-VEGF therapy.

Here, we analyzed nine SNPs in autophagy-related genes (Table 1) for the association with nAMD in a cohort of Russian patients. Further, we studied the effect of these gene variants on anatomical and functional response to aflibercept treatment.

Table 1. SNPs analyzed in the study.

Gene	SNP	Location/Consequence	Position	Minor Allele	MAF
ATG5	rs573775	Intron variant	chr6:106316991	A	0.27
MAP1LC3A	rs73105013	Intron variant	chr20:34557008	C	0.08
MTOR	rs1064261	Missense variant	chr1:11228701	G	0.28
	rs1057079	Synonymous variant	chr1:11145001	C	0.26
	rs11121704	Intron variant	chr1:11233902	C	0.28
	rs2295080	Upstream variant	chr1:11262571	G	0.31
SQSTM1	rs10277	3 Prime UTR Variant	chr5:179837731	T	0.48
ULK1	rs11246867	Upstream variant	chr12:131893472	A	0.06
	rs3088051	3 Prime UTR Variant	chr12:131922463	C	0.3

SNP: single nucleotide polymorphism. MAF: minor allele frequency in European populations according to 1000 Genomes.

2. Materials and Methods

2.1. Study Participants

This study was conducted in accordance with the ethical principles of the Declaration of Helsinki and the National Standard for Good Clinical Practice and was approved by the Institutional Review Board at the Institute of Cytology and Genetics SB RAS. The subjects were informed and had provided written consent to the collection and scientific use of the specimen prior to the procedure. The AMD group consisted of 315 patients (98 (31%) men and 217 (69%) women with a mean age of 71.5 ± 8.6 years) diagnosed with nAMD at the Department of Ophthalmology of the Novosibirsk Regional Clinical Hospital. The control group comprised 317 subjects (103 (32%) men and 214 (68%) women with a mean age of 66.5 ± 7.5 years), undergoing routine cataract surgery without a history of AMD and macular changes such as drusen or pigment abnormalities [22]. A complete ophthalmological examination was performed, including visometry, biomicroscopy, ophthalmoscopy, and optical coherence tomography (OCT). Exclusion criteria from the study were active neovascularization in the periphery of the retina and in the anterior segment of the eye, a history of laser photocoagulation, medical intravitreal therapy in history, spherical equivalent more than ±6.0 diopters, uveitis, geographic atrophy, surgical interventions on the vitreous body, the presence of signs of intraocular inflammation, pathology of the vitreomacular interface with traction component, polypoidal choroidal vasculopathy or any other confounding retinopathies [12,22].

2.2. Optical Coherence Tomography Study

The subset OCT study included 196 treatment-naive patients diagnosed with nAMD. After being diagnosed, all patients began to receive anti-VEGF therapy. Intravitreal injections (IVIs) of aflibercept (Regeneron, Munich, Germany) (0.05 mL (2 mg)) were administered according to the standard method in the operating room after local epibulbar anesthesia with an alkaline solution (Alcon, Fort Worth, TX, USA) through a 31 G needle at least 3 mm from the limbus. Three successive injections were performed with an interval of 4 weeks. To assess the effectiveness of therapy, a clinical and instrumental examination of patients was performed by OCT (Cirrus HD-OCT, Humphrey Zeiss, Inc., Jena, Germany) and visometry, with the determination of best corrected visual acuity (BCVA) at baseline and after three IVIs [22]. The following parameters were assessed: type of macular neovascularization (MNV), BCVA, central retinal thickness (CRT), height of pigment epithelium detachment (PED), height of subretinal fluid (SRF) and the presence of intraretinal fluid (IRF). BCVA was estimated using a letter count on the Early Treatment of Diabetic Retinopathy Study (ETDRS) chart.

2.3. DNA Isolation and Genotyping

Peripheral venous blood was collected in vacutainers with EDTA for DNA analysis at baseline visit. Genomic DNA was isolated by DNA Blood Kit (Biolabmix, Novosibirsk, Russia) according to the manufacturer's protocol. Genotyping was carried out with TaqMan-based allelic discrimination assays. Primers and probes were designed using Primer-Blast (https://www.ncbi.nlm.nih.gov/tools/primer-blast/ (accessed on 1 January 2021) and Oligo Analyzer (version 1.0.3) (Table 2). LNA (locked nuckeic acid) modifications were used to obtain the optimal melting temperature in probes. PCR was performed in 20 μL reaction volume containing 20 ng of genomic DNA, BioMaster HS-qPCR (2×) buffer (Biolabmix), 0.3 mM primers and 0.1 mM FAM/VIC-conjugated probes. PCR thermal cycling conditions were as follows: denaturation for 3 min at 95 °C followed by 35 cycles, including denaturation at 95 °C for 10 s, primer annealing and subsequent elongation at 60 °C for 30 s. Amplification was conducted using CFX96 Thermal Cycler (Bio-Rad, Hercules, CA, USA). The PCR data were processed using «Bio-Rad CFX Manager 3.1» software, Russian Edition #1845028, Novosibirsk, Russia. To verify the results of allelic discrimination, Sanger sequencing was used in samples from different genotypes on an ABI 3500 DNA sequencer (Thermo Fisher Scientific, Waltham, MA, USA) by means of the BigDye Terminator v3.1 Cycle Sequencing Kit (Thermo Fisher Scientific, Waltham, MA, USA).

Table 2. Primers/probes sequences.

rsID	Gene Name	Sequence
rs573775	*ATG5*	Forward 5′-CCCTACCTAGTATGCTCCTC-3′ Reverse 5′-AAAAGCCATGTCCTTATGCC-3′ 5′-FAM-CCTCTGGCCCCAGTGAAACAG-BHQ1-3′ 5′-VIC-CTCTGGCCCCA[+A]TGAAACAGT-BHQ1-3′
rs1064261	*MTOR*	Forward 5′-AAGGATTGGGGTTTGAGGTA-3′ Reverse 5′-GACCAGTCACTCTCTCATCA-3′ 5′-FAM-CACGTTCCTTAA[+C]GTCATTCGA-BHQ1-3′ 5′-VIC-CACGTTCCTTAA[+T]GTCATTCGA-BHQ1-3′
rs1057079	*MTOR*	Forward 5′-GCAGCCTGTAAGTTCTCAAT-3′ Reverse 5′-CCCAAGGGTTGTTTCTCTTC-3′ 5′-FAM-CTCCTGCCATCGCAGTTAATTCA-BHQ1-3′ 5′-VIC-CTCCTGCCAT[+T]GCAGTTAATT-BHQ1-3′
rs11121704	*MTOR*	Forward 5′-TTTTTCCTCATTTTGGGCGA-3′ Reverse 5′-TATCAGTTGCAGGAAAGTGC-3′ 5′-FAM-CAGGCACATCATCGCAGATGTTT-BHQ1-3′ 5′-VIC-CAGGCACATCATCACAGATGTTTG-BHQ1-3′
rs2295080	*MTOR*	Forward 5′-TTCCCCGCTGTCCTCTA-3′ Reverse 5′-GCCTGTTTTTCAGTCCATCT-3′ 5′-FAM-CCTCAGGGCTGGGAACCC-BHQ1-3′ 5′-VIC-CCTCAGGG[+A]TGGGAACCCTC-BHQ1-3′
rs73105013	*MAP1LC3A*	Forward 5′-CAGCCTTAAAAACAAAAACCCT-3′ Reverse 5′-ATGGAAGGCAGAAAGGGAGA-3′ 5′-FAM-CTTATCCCCAG[+T]GTCTTCTGC-BHQ1-3′ 5′-VIC-CTTATCCCCAG[+C]GTCTTCTGC-BHQ1-3′
rs10277	*SQSTM1*	Forward 5′-GTCCCTCTGAAGAGACCTTG-3′ Reverse 5′-CTGGGAAGGAGCTATGGAG-3′ 5′-FAM-AGGACAAAT[+T]GCGCCCAT-BHQ1-3′ 5′-VIC-CAGGACAAATCGCGCCCATT-BHQ1-3′
rs11246867	*ULK1*	Forward 5′-GTACGGTGAACAGCACTAAC-3′ Reverse 5′-CAGCCAAAAGAGCCCG-3′ 5′-FAM-CAGCCAACAG[+C]GATTGCTCT-BHQ1-3′ 5′-VIC-CAGCCAACAG[+T]GATTGCTCT-BHQ1-3′
rs3088051	*ULK1*	Forward 5′-GGAAGCAGATGAGGGGAATA-3′ Reverse 5′-CTCTCTGCAGATGCCCTC-3′ 5′-FAM-CAGTCAGTTT[+T]GATGTCAGCTC-BHQ1-3′ 5′-VIC-CAGTCAGTTT[+C]GATGTCAGCTC-BHQ1-3′

[+X]—LNA modifications.

2.4. Statistical Analysis

An analysis of the comparisons of frequencies of genotypes between AMD and control groups was performed using the chi-square test. SNPs with genotype frequencies that differ significantly between groups (p-value < 0.05, $\chi^2 < 5.991$) were selected for further analysis. To evaluate the effects of these SNPs, odds ratios (ORs) and 95% confidence intervals (CIs) were calculated using a logistic regression analysis adjusted for sex and age adopting codominant, dominant, recessive, overdominant and additive models of inheritance using SNPstats [23]. The significance threshold after the implementation of Bonferroni correction for multiple testing was set on $p = 0.05/15 = 0.003$.

Statistical analysis of association between SNPs and OCT markers was performed using StatTech v. 3.1.10 (Developer—StatTech LLC, Russia). Quantitative variables were assessed for normality using the Kolmogorov–Smirnov test. Quantitative variables following a normal distribution were described using mean (M) and standard deviation (SD). Quantitative variables following non-normal distribution were described using median (Me) and lower and upper quartiles (Q1–Q3). Categorical data are shown as absolute values (percentage). Comparisons of three groups on a quantitative variable whose distribution differed from normal were made using the Kruskal–Wallis test and Dunn's criterion with Holm correction as a post hoc method. Comparison of frequencies in the analysis of multifield contingency tables was performed using Pearson's chi-square test (for expected values greater than 10). Wilcoxon test was used for comparison of quantitative variable following non-normal distribution between two matched samples. Comparison of binary variables in two paired samples was performed using McNemar test. The differences were considered significant at $p < 0.05$.

3. Results

3.1. Association with Risk of nAMD

The genotypes of SNPs of the mTOR (rs1064261, rs1057079, rs11121704, rs2295080), SQSTM1 (rs10277), Ulk1 (rs11246867, rs3088051), MAP1LC3A (rs73105013) and Atg5 (rs573775) genes were determined in the AMD and in the control group (Table 3). The allele frequencies calculated for ethnic Russians in our study were close or similar to the corresponding frequencies in European populations reported by the 1000 Genomes Project. The genotype frequencies of rs1057079, rs11121704 and rs2295080 of the mTOR gene differ significantly between groups (p-value < 0.05, $\chi^2 < 5.991$). These SNPs were selected for an logistic regression analysis adjusted for sex and age to evaluate the odds ratios (ORs) and 95% confidence intervals (CIs) (Table 4).

Table 3. Allele and genotype frequencies in the nAMD and control groups in a Russian cohort.

SNP	Genotype/Allele	Control	AMD	p-Value, χ^2
rs573775 *ATG5*	A/A	24 (8%)	26 (8%)	$p = 0.838$, 0.354
	G/A	145 (46%)	137 (43%)	
	G/G	148 (47%)	152 (48%)	
	MAF	0.3	0.3	
rs1064261 *MTOR*	A/A	150 (47%)	125 (40%)	$p = 0.093$, 4.757
	A/G	130 (41%)	156 (50%)	
	G/G	37 (12%)	34 (11%)	
	MAF	0.32	0.36	
rs1057079 *MTOR*	C/C	28 (9%)	21 (7%)	$p = 0.002$, 12.563
	T/C	109 (34%)	152 (48%)	
	T/T	180 (57%)	142 (45%)	
	MAF	0.26	0.31	

Table 3. *Cont.*

SNP	Genotype/Allele	Control	AMD	*p*-Value, χ^2
rs11121704 *MTOR*	C/C	28 (9%)	29 (9%)	p = 0.006, 10.324
	T/C	121 (38%%)	158 (50%)	
	T/T	168 (53%)	128 (41%)	
	MAF	0.28	0.34	
rs2295080 *MTOR*	G/G	36 (11%)	31(10%)	p = 0.002, 12.562
	T/G	123 (39%)	166 (53%)	
	T/T	158 (5%)	118 (37%)	
	MAF	0.31	0.36	
rs73105013 *MAP1LC3A*	C/C	2 (0.6%)	0 (0%)	p = 0.169, 3.557
	T/C	41 (13%)	52 (17%)	
	T/T	273 (86%)	261 (83%)	
	MAF	0.07	0.08	
rs10277 *SQSTM1*	C/C	118 (37%)	106 (34%)	p = 0.385, 1.913
	C/T	157 (50%)	156 (50%)	
	T/T	42 (13%)	53 (17%)	
	MAF	0.38	0.42	
rs11246867 *ULK1*	A/A	2 (0.6%)	1 (0.3%)	p = 0.574, 1.113
	G/A	33 (10%)	40 (13%)	
	G/G	282 (89%)	274 (87%)	
	MAF	0.06	0.07	
rs3088051 *ULK1*	C/C	32 (10%)	27 (9%)	p = 0.579, 1.096
	T/C	121 (38%)	132 (42%)	
	T/T	164 (52%)	156 (50%)	
	MAF	0.29	0.3	

χ^2—chi-square test. The critical value of χ^2 at the significance level $p < 0.05$ is 5.991. The null hypothesis is rejected if $\chi^2 > 5.99$.

Table 4. Association of the *MTOR* polymorphisms with the risk of nAMD in a Russian cohort.

SNP	Model of Inheritance	OR (95% CI) Adjusted for Sex and Age by Logistic Regression	*p*-Value	AIC
rs1057079 *MTOR*	Codominant:			
	C/T vs. T/T	1.85 (1.31–2.62)	0.0018	814.5
	C/C vs. T/T	1.07 (0.57–2.03)		
	Dominant: C/T-C/C vs. T/T	1.70 (1.22–2.36)	0.0017	815.2
	Overdominant: C/T vs. C/C-T/T	1.83 (1.31–2.56)	0.0004	812.5
	Recessive: C/C vs. C/T-T/T	0.82 (0.44–1.51)	0.51	824.6
	Additive	1.34 (1.03–1.74)	0.028	820.2
rs11121704 *MTOR*	Codominant:			
	C/T vs. T/T	1.74 (1.23–2.46)	0.0067	817
	C/C vs. T/T	1.47 (0.81–2.68)		
	Dominant: C/T-C/C vs. T/T	1.69 (1.21–2.35)	0.0018	815.3
	Overdominant: C/T vs. C/C-T/T	1.63 (1.17–2.27)	0.0038	816.7
	Recessive: C/C vs. C/T-T/T	1.13 (0.64–2.00)	0.68	824.9
	Additive	1.40 (1.09–1.81)	0.0094	818.3
rs2295080 *MTOR*	Codominant:			
	G/T vs. T/T	1.86 (1.31–2.64)	0.0021	814.8
	G/G vs. T/T	1.29 (0.73–2.26)		
	Dominant: G/T-G/G vs. T/T	1.74 (1.24–2.42)	0.0011	814.4
	Overdominant: G/T vs. G/G-T/T	1.77 (1.27–2.47)	0.0007	813.5
	Recessive: G/G vs. G/T-T/T	0.93 (0.55–1.59)	0.8	825
	Additive	1.35 (1.05–1.73)	0.019	819.6

OR: odds ratio; CI: confidence interval; AIC: Akaike's information criterion. The significance threshold taking into account multiple comparisons is $p = 0.003$.

The MTOR rs1057079 C/T genotype was associated with increased odds of nAMD risk under the codominant (OR = 1.85; CI: 1.31–2.61; $p = 0.0018$) and overdominant (OR = 1.83; CI: 1.31–2.56; $p = 0.00004$) models, and C/T-C/C genotypes were associated with 1.7-fold (OR = 1.7; CI: 1.22–2.36; $p = 0.0017$) increased odds of nAMD according to the dominant model (Table 4). Each C allele increases the odds of developing nAMD by 1.34-fold under the additive model (OR = 1.54; CI: 1.07–2.20; $p = 0.017$). According to Akaike's information criterion (AIC), the overdominant model was preferable.

The MTOR rs11121704 C/T genotype was associated with increased odds of developing AMD under the codominant (OR = 1.74; CI: 1.23–2.46; $p = 0.0067$) and overdominant (OR = 1.63; CI: 1.17–2.27; $p = 0.0038$) models, and C/T+C/C genotypes were associated with 1.69-fold (OR = 1.69; CI: 1.21–2.35; $p = 0.0018$) increased odds of AMD according to the dominant model. Each C allele increases the odds of developing AMD by 1.4-fold under the additive model (OR = 1.4; CI: 1.09–1.81; $p = 0.0094$). According to AIC, the dominant model was preferable.

The MTOR rs2295080 G/T genotype was associated with increased odds of a risk of nAMD under the codominant (OR = 1.86; CI: 1.31–2.64; $p = 0.0021$) and overdominant (OR = 1.77; CI: 1.27–2.47; $p = 0.0007$) models, and G/T+G/G genotypes were associated with 1.74-fold (OR = 1.74; CI: 1.24–2.42; $p = 0.0011$) increased odds of AMD according to the dominant model. Each G allele increases the odds of developing AMD by 1.35-fold under the additive model (OR = 1.35; CI: 1.05–1.73; $p = 0.019$). According to AIC, the overdominant model was preferable.

Bonferroni correction was performed to reduce type I error in multiple testing, and a significant threshold was set at 0.05/15 = 0.003 for genotype analyses. After that, only the codominant, dominant and overdominant models for rs1057079; the dominant model for rs11121704; and the dominant and overdominant models for rs2295080 were still valid.

The overdominant model of inheritance may be explained with the fact that all components, including mTOR, are dimerized in the mTORC1 and mTORC2 complexes. Non-coding substitutions in the MTOR gene can lead to changes in mRNA stability, interaction with transcription factors, and the rate of protein synthesis, and the simultaneous presence of two heterogeneous protein populations in heterozygous cells can affect the structure or concentration of an efficient homodimer in the cell.

An analysis of linkage disequilibrium (LD) shows strong non-random association of all variants in the MTOR gene (spanning region of 117.5 kb) with D′ > 0.867. The GCCG haplotype, corresponding to the minor alleles of rs1064261, rs1057079, rs11121704 and rs2295080, respectively, was associated with 1.45-fold (OR = 1.54; CI: 1.09–1.92; $p = 0.01$) increased odds of nAMD compared with the most frequent ATTT haplotype.

No evidence of association with nAMD risk was observed for the other SNPs.

3.2. Association between SNPs and OCT Markers

In total, 196 patients with treatment-naive nAMD were included in the OCT study. The mean age of the study population was 71 ± 9 years, of which there were 60 (30.6%) males and 136 (69.4%) females. The OCT characteristics of the study participants are shown in Table 5. According to the type of MNV, MNV type 1 was diagnosed in 93 (47.4%) eyes, MNV type 2 was diagnosed in 94 (48%) eyes and MNV type 3 was diagnosed in 9 (4.6%) eyes. The median baseline PED height was 126 μm (interquartile range (IQR) 89 μm to 190 μm), mean baseline CRT was 316 μm (IQR 271 μm to 372 μm) and median baseline SRF height was 126 μm (IQR 89 μm to 190 μm). IRF was seen in 140 eyes (71.4%). According to OCT, a significant BCVA gain and significant decreases in CRT, PED and SRF height were achieved in patients after three loading IVIs (Table 5).

Table 5. Characterization of functional and anatomical parameters of the retina in patients with nAMD.

	Baseline	3 IVIs	p-Value
BCVA, letters	48 ± 22	55 ± 21	<0.001 [a]
CRT, mkm	316 [271–372]	246 [218–289]	<0.001 [a]
PED height, mkm	126 [89–190]	46 [21–96]	<0.001 [a]
SRF height, mkm	56 [29–90]	22 [0–44]	<0.001 [a]
IRF, abs. (%)	140 (71.4)	89 (45.4)	<0.001 [b]

BCVA—best corrected visual acuity; CRT—central retinal thickness; PED—detachment of the pigment epithelium; SRF—subretinal fluid; IRF—intraretinal fluid. Mean ± sd or Median (Q1–Q3). Applied methods for matched samples: [a] Wilcoxon test, [b] McNemar test.

The decrease in CRT was associated with SQSTM1-rs10277 ($p = 0.001$): it was significantly higher in TT (93 [58–122] μm) and CT (66 [30–105] μm) carriers compared with the CC genotype (47 [24–68] μm) (Table 6).

Table 6. Analysis of decrease in CRT depending on *SQSTM1*-rs10277.

SNP	Genotype	Decrease in CRT, μm			p-Value
		Me	Q1–Q3	n	
rs10277 *SQSTM1*	C/C	47	24–68	63	0.001
	C/T	66	30–105	100	$p_{C/T–C/C} = 0.014$
	T/T	93	58–122	33	$p_{T/T–C/C} = 0.002$

Applied method: Kruskal–Wallis test and Dunn's criterion with Holm correction as a post hoc method.

The BCVA was increased in 117 (59.7%) patients by 10 [5–20] letters, did not changed in 59 (30.1%) and was decreased in 20 (10.2%) patients. ULK1-rs3088051 was associated with BCVA change. Among patients with the TT and CT genotypes for the ULK1-rs3088051, an improvement in visual acuity was noted in 67.6% and 53.8% of cases, while in patients with the CC genotype, an increase in BCVA was recorded in 37.5% of cases ($p = 0.01$) (Table 7, Figure 1). Moreover, in carriers of the CC genotype, the median BCVA change was the smallest—0 letters (IQR −0 to −6 letters)—while in CT and TT carriers, it was 5 letters (IQR 0 to 12 letters) and 5 letters (IQR 0 to 16 letters), respectfully ($p = 0.008$, Table 7).

Table 7. Analysis of BCVA change after three IVIs depending on *ULK1*-rs3088051.

***ULK1*-rs3088051**	**BCVA Change, abs. (%).**			**p-Value, Pearson's Chi-Square**
	Decline	**no Change**	**Increase**	
C/C	4 (25.0)	6 (37.5)	6 (37.5)	0.013
C/T	12 (15.4)	24 (30.8)	42 (53.8)	$p_{C/C–T/T} = 0.010$
T/T	4 (3.9)	29 (28.4)	69 (67.6)	$p_{C/T–T/T} = 0.037$
ULK1-rs3088051	BCVA change, letters.			p-value, Kruskal–Wallis
	Me	Q1–Q3	n	
C/C	0	−0–6	16	0.008
C/T	5	0–12	78	$p_{T/T–C/C} = 0.022$
T/T	5	0–16	102	

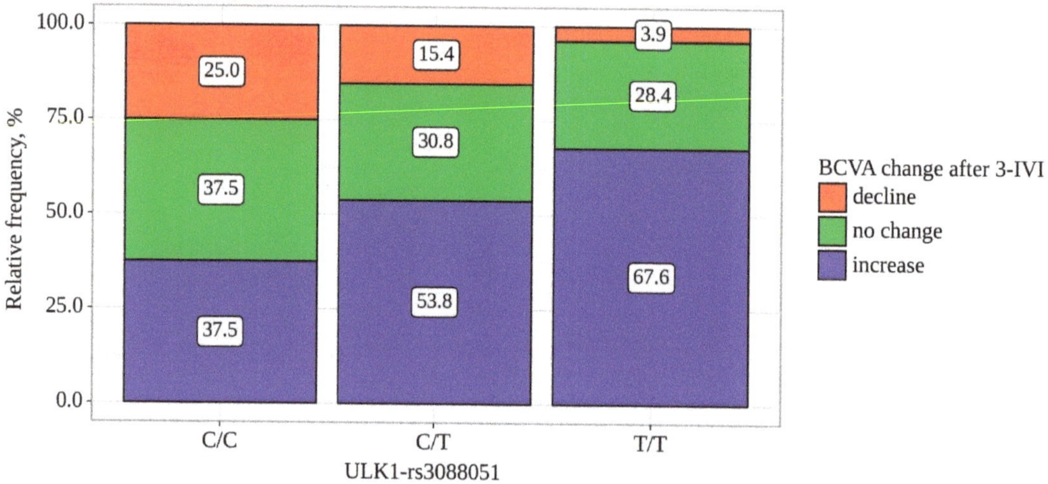

Figure 1. Analysis of BCVA change after three IVIs depending on *ULK1*-rs3088051.

Other SNPs did not show significant associations with the outcome of anti-VEGF treatment.

4. Discussion

AMD is a complex degenerative disease of the retina with multiple risk-modifying factors, including aging, genetics and diet. The progression of AMD is initially characterized by atrophic alterations in the RPE, as well as the formation of lysosomal lipofuscin and extracellular drusen deposits [21]. Autophagy is a catabolic process indispensable for retinal cell homeostasis. The role of autophagy in AMD pathology and treatment is steadily emerging [24]. Although anti-VEGF therapy has achieved a good therapeutic effect in nAMD, the recurrence of MNV is inevitable, which increases the cost of treatment [25]. Therefore, it is especially important to uncover the mechanisms of the processes that control the development of MNV and to reduce its recurrence. According to the latest data [17,18,21], autophagy may be one of these controlling processes. In particular, attention has been drawn to research in the field of cancer therapy, where, as in nAMD, anti-VEGF drugs are used [26]. Recent studies have shown that autophagy plays a key role in the mechanism of formation of tumor resistance to antiangiogenic therapy [17].

As well as in tumor treatment, autophagy may be responsible for resistance development to anti-VEGF therapy during nAMD treatment. Recently, the in vitro study showed that ranibizumab and conbercept can trigger the autophagy of vascular endothelial cells while aflibercept can inhibit it. The mechanism of autophagy activation was related to the activation of the p53/DRAM pathway [25].

In this study, we showed that regulatory SNPs (rs1057079, rs11121704 and rs2295080) in the MTOR gene are moderately associated with a risk of nAMD in the Russian population. The kinase mTOR (mechanistic Target of Rapamycin) affects the most fundamental cellular processes. It is essential for cellular and organismal physiology, and its dysregulation is frequently associated with human aging and age-related diseases. As the PI3K/AKT/mTOR signaling pathway's key player, mTOR is essential for the sensing of cellular energy, oxygen and nutrients [27]. The most significant stimulus that controls mTORC1 activity is nutrient deficiency. Actually, mTORC1 serves as an amino acid molecular sensor, connecting cellular demand with dietary supply [28]. Acting as a crucial upstream regulator of autophagy, mTOR inhibits the ULK1-ATG13-RB1CC1/FIP200 complex on the initial step of autophagy. Recently, it was shown that the mTOR pathway is partially responsible for the mitochondrial damage caused by complement factor H protein loss in RPE cells [29]. Previously, a similar link between SNPs (rs1057079, rs11121704 and rs2295080) in the MTOR gene and wet AMD was found in the Finnish population [16]. It is known that pharmacogenetic association may differ by ethnicity if the allele distribution of a candidate polymorphism varies among populations. As a result, the studies aimed at identifying or replicating genes for susceptibility to AMD risk and response to treatment in various populations continue to be relevant [12].

The main aim of pharmacogenetic studies is to identify the patients who could benefit the most from treatment. Our study links polymorphisms in autophagy-related genes to anti-VEGF treatment response. We showed that Sqstm1-rs10277 and Ulk1-rs3088051 influence short-term response to aflibercept therapy. The results are consistent with the study by Paterno et al. [16], where a correlation was also found between these polymorphisms and the response to antiangiogenic therapy.

The members of the ULK (Unc-51-like kinase) family of proteins are the orthologues of the yeast Atg1, a serine/threonine protein kinase essential for autophagy initiation [30]. ULK1 or ULK2 are part of a protein complex that is responsible for driving autophagy initiation upon autophagy-inducing stimuli [31]. Moreover, the ULK1/2 protein complex carries other different autophagy-related functions, such as ATG9-vesicle recruitment or regulation of ATG4B activity, and contributes to regulating the mitophagy and degradation of protein aggregates [32]. Several pathogenic variants of the ULK1/2 kinase complex have been identified [30]. Polymorphisms in members of the ULK1/2 kinase complex have been associated with a variety of pathologies related to immune system dysfunction: Crohn's disease susceptibility [33], tuberculosis [34] and ankylosing spondylitis [35].

The SQSTM1 gene codes a p62 protein—a multifunctional protein important for protein aggregate degradation, mitophagy and the engulfment of intracellular pathogens by autophagosomes [36] and the most studied ubiquitin-binding selective autophagy receptor (SAR). SARs mediate the recognition and engulfment of specific cargo in autophagosomes by simultaneously binding both to the target molecules and to the ATG8 proteins conjugated on the concave side of the autophagosomal membrane [37]. The dysregulation of SAR function plays a role in the pathogenesis of a wide variety of diseases, with most pathogenic variants being identified in the SQSTM1 gene [30]. Nucleotide changes on SQSTM1 are known to contribute to the origin of neurological alterations [38], implicated in the pathogenesis of amyotrophic lateral sclerosis [39] and frontotemporal dementia [40].

Obviously, angiogenesis, including VEGF-dependent pathological vascular growth, and autophagy are closely related processes. Our results fit into the concept recently proposed by the Grosjean et al. [21]. The authors of this model proposed the term "autophagopathy" to encompass a class of genetic diseases whose etiology is associated with a defect in the autophagy machinery, whether directly related to abnormal autophagic flux, LC3-associated phagocytosis or any concomitant process of cellular debris removal. This model suggests that neither genetic variants of autophagy nor environmental exposures alone cause disease. However, SNPs that result in low-level autophagy may alter the cell's ability to detoxify damaged organelles when exposed to environmental factors. As a consequence, regulatory autophagy-related SNPs will predispose individuals to develop broad range of diseases (such as cancer, infections, neurodegenerative, cardiovascular, metabolic and inflammatory diseases) only when exposed to this specific environmental risk, the nature of which will determine the affected organ and disease [21]. Therefore, it will be critical to recognize these autophagy-related SNP carriers to aid disease screening, prevention and precision treatment programs to rapidly alleviate comorbid complications [21]. Similarly, it can be assumed that the carriage of certain variants of autophagy genes, which determine the characteristics of the autophagy apparatus under stress, can influence the success of antiangiogenic therapy, as we and Paterno et al. [16] have shown for nAMD. Therefore, autophagy SNP genotyping may be useful for investigating treatment response to other diseases that use an antiangiogenic protocol, such as diabetic macular edema, retinopathy of prematurity and various types of cancer. To our knowledge, there are no ongoing clinical trials targeted at the autophagy pathway in nAMD. Our results suggest that such trials may be relevant and that autophagy could be a target for future drugs to overcome resistance to anti-VEGF therapy.

A limitation of our study is the relatively small sample size of the patient cohort, which reduces the statistical power available to detect statistically significant associations. Another limitation of our study is the use of patients with cataracts as a control group due to the inability to recruit healthy elderly people. Also, we studied only short-term response. On the other hand, a significant time point for the prediction of long-term functional outcomes may be the period after the three IVIs associated with the loading phase [41]. The strength of this study is the inclusion of patients who had never received treatment, thereby eliminating any potential influence of a retreatment regimen.

5. Conclusions

Our study provides further insight into the pharmacogenetics of the clinical response of nAMD to anti-VEGF therapy by identifying an association between autophagy-related gene polymorphisms and treatment response outcome. We confirmed the previously found effects of polymorphisms in autophagy genes on both nAMD risk and anti-VEGF response in a larger and independent cohort of patients. Based on our results, (1) variants in the mTOR gene are associated with nAMD risk, and (2) Sqstm1-rs10277 and Ulk1-rs3088051 could be considered as biomarkers of response to anti-VEGF treatment in nAMD patients. However, to uncover the molecular mechanisms mediating the influence of autophagy on VEGF pathway in AMD vulnerable eye tissues, further studies with the inclusion of in vivo and in vitro model systems are necessary.

Author Contributions: Conceptualization, O.S.K.; investigation, A.S.D., I.F.N., V.A.D., O.S.K. and A.Z.F.; data curation, A.Z.F. and O.S.K.; writing—original draft preparation, O.S.K.; writing—review and editing, A.Z.F. and N.G.K.; supervision, A.Z.F. and N.G.K.; project administration, O.S.K.; funding acquisition, O.S.K. All authors have read and agreed to the published version of the manuscript.

Funding: This research was funded by the Russian Science Foundation (grant # 21-15-00047).

Institutional Review Board Statement: The study was conducted in accordance with the ethical principles of the Declaration of Helsinki and the National Standard for Good Clinical Practice, and approved by the Institutional Review Board of the Institute of Cytology and Genetics, SB RAS (protocol No. 127, date of approval 13 January 2017).

Informed Consent Statement: Informed consent was obtained from all subjects involved in the study.

Data Availability Statement: The datasets are available from the corresponding author on request.

Acknowledgments: Genotyping was performed at the Microscopy Center of the ICG SB RAS, Novosibirsk, Russia.

Conflicts of Interest: The authors declare no conflict of interest.

References

1. Li, J.Q.; Welchowski, T.; Schmid, M.; Mauschitz, M.M.; Holz, F.G.; Finger, R.P. Prevalence and incidence of age-related macular degeneration in Europe: A systematic review and meta-analysis. *Br. J. Ophthalmol.* **2020**, *104*, 1077–1084. [CrossRef]
2. Mitchell, P.; Liew, G.; Gopinath, B.; Wong, T.Y. Age-related macular degeneration. *Lancet* **2018**, *392*, 1147–1159. [CrossRef]
3. Fursova, A.Z.; Derbeneva, A.S.; Vasilyeva, M.A.; Nikulich, I.F.; Tarasov, M.S.; Gamza, Y.A.; Chubar, N.V.; Gusarevitch, O.G.; Dmitrieva, E.I.; Kozhevnikova, O.S.; et al. New findings on pathogenetic mechanisms in the development of age-related macular degeneration. *Vestn. Oftalmol.* **2022**, *138*, 120–130. [CrossRef]
4. Spaide, R.F.; Jaffe, G.J.; Sarraf, D.; Freund, K.B.; Sadda, S.R.; Staurenghi, G.; Waheed, N.K.; Chakravarthy, U.; Rosenfeld, P.J.; Holz, F.G.; et al. Consensus Nomenclature for Reporting Neovascular Age-Related Macular Degeneration Data: Consensus on Neovascular Age-Related Macular Degeneration Nomenclature Study Group. *Ophthalmology* **2020**, *127*, 616–636. [CrossRef]
5. Nguyen, D.D.; Luo, L.J.; Yang, C.J.; Lai, J.Y. Highly Retina-Permeating and Long-Acting Resveratrol/Metformin Nanotherapeutics for Enhanced Treatment of Macular Degeneration. *ACS Nano* **2023**, *17*, 168–183. [CrossRef]
6. Cobos, E.; Recalde, S.; Anter, J.; Hernandez-Sanchez, M.; Barreales, C.; Olavarrieta, L.; Valverde, A.; Suarez-Figueroa, M.; Cruz, F.; Abraldes, M.; et al. Association between CFH, CFB, ARMS2, SERPINF1, VEGFR1 and VEGF polymorphisms and anatomical and functional response to ranibizumab treatment in neovascular age-related macular degeneration. *Acta Ophthalmol.* **2018**, *96*, e201–e212. [CrossRef]

7. Yang, S.; Zhao, J.; Sun, X. Resistance to anti-VEGF therapy in neovascular age-related macular degeneration: A comprehensive review. *Drug Des. Dev. Ther.* **2016**, *10*, 1857–1867. [CrossRef]
8. Lee, A.Y.; Raya, A.K.; Kymes, S.M.; Shiels, A.; Brantley, M.A., Jr. Pharmacogenetics of complement factor H (Y402H) and treatment of exudative age-related macular degeneration with ranibizumab. *Br. J. Ophthalmol.* **2009**, *93*, 610–613. [CrossRef]
9. Hong, N.; Shen, Y.; Yu, C.Y.; Wang, S.Q.; Tong, J.P. Association of the polymorphism Y402H in the CFH gene with response to anti-VEGF treatment in age-related macular degeneration: A systematic review and meta-analysis. *Acta Ophthalmol.* **2016**, *94*, 334–345. [CrossRef]
10. Dedania, V.S.; Grob, S.; Zhang, K.; Bakri, S.J. Pharmacogenomics of response to anti-VEGF therapy in exudative age-related macular degeneration. *Retina* **2015**, *35*, 381–391. [CrossRef]
11. Medina, F.; Motta, A.; Takahashi, W.Y.; Carricondo, P.C.; Motta, M.; Melo, M.B.; Vasconcellos, J. Association of the CFH Y402H Polymorphism with the 1-Year Response of Exudative AMD to Intravitreal Anti-VEGF Treatment in the Brazilian Population. *Ophthalmic Res.* **2019**, *61*, 168–173. [CrossRef] [PubMed]
12. Kozhevnikova, O.S.; Fursova, A.Z.; Derbeneva, A.S.; Nikulich, I.F.; Tarasov, M.S.; Devyatkin, V.A.; Rumyantseva, Y.V.; Telegina, D.V.; Kolosova, N.G. Association between Polymorphisms in CFH, ARMS2, CFI, and C3 Genes and Response to Anti-VEGF Treatment in Neovascular Age-Related Macular Degeneration. *Biomedicines* **2022**, *10*, 1658. [CrossRef] [PubMed]
13. Akiyama, M.; Takahashi, A.; Momozawa, Y.; Arakawa, S.; Miya, F.; Tsunoda, T.; Ashikawa, K.; Oshima, Y.; Yasuda, M.; Yoshida, S.; et al. Genome-wide association study suggests four variants influencing outcomes with ranibizumab therapy in exudative age-related macular degeneration. *J. Hum. Genet.* **2018**, *63*, 1083–1091. [CrossRef]
14. Balikova, I.; Postelmans, L.; Pasteels, B.; Coquelet, P.; Catherine, J.; Efendic, A.; Hosoda, Y.; Miyake, M.; Yamashiro, K.; ANGEL Study Group Members; et al. Genetic biomarkers in the VEGF pathway predicting response to anti-VEGF therapy in age-related macular degeneration. *BMJ Open Ophthalmol.* **2019**, *4*, e000273. [CrossRef]
15. Wu, M.; Xiong, H.; Xu, Y.; Xiong, X.; Zou, H.; Zheng, M.; Wang, X.; Zhou, X. Association between VEGF-A and VEGFR-2 polymorphisms and response to treatment of neovascular AMD with anti-VEGF agents: A meta-analysis. *Br. J. Ophthalmol.* **2017**, *101*, 976–984. [CrossRef]
16. Paterno, J.J.; Koskela, A.; Hyttinen, J.M.T.; Vattulainen, E.; Synowiec, E.; Tuuminen, R.; Watala, C.; Blasiak, J.; Kaarniranta, K. Autophagy Genes for Wet Age-Related Macular Degeneration in a Finnish Case-Control Study. *Genes* **2020**, *11*, 1318. [CrossRef] [PubMed]
17. Chandra, A.; Rick, J.; Yagnik, G.; Aghi, M.K. Autophagy as a mechanism for anti-angiogenic therapy resistance. *Semin Cancer Biol.* **2020**, *66*, 75–88. [CrossRef]
18. Kocak, M.; Ezazi Erdi, S.; Jorba, G.; Maestro, I.; Farres, J.; Kirkin, V.; Martinez, A.; Pless, O. Targeting autophagy in disease: Established and new strategies. *Autophagy* **2022**, *18*, 473–495. [CrossRef]
19. Kozhevnikova, O.S.; Telegina, D.V.; Tyumentsev, M.A.; Kolosova, N.G. Disruptions of Autophagy in the Rat Retina with Age During the Development of Age-Related-Macular-Degeneration-like Retinopathy. *Int. J. Mol. Sci.* **2019**, *20*, 4804. [CrossRef]
20. Blasiak, J.; Pawlowska, E.; Szczepanska, J.; Kaarniranta, K. Interplay between Autophagy and the Ubiquitin-Proteasome System and Its Role in the Pathogenesis of Age-Related Macular Degeneration. *Int. J. Mol. Sci.* **2019**, *20*, 210. [CrossRef]
21. Grosjean, I.; Romeo, B.; Domdom, M.A.; Belaid, A.; D'Andrea, G.; Guillot, N.; Gherardi, R.K.; Gal, J.; Milano, G.; Marquette, C.H.; et al. Autophagopathies: From autophagy gene polymorphisms to precision medicine for human diseases. *Autophagy* **2022**, *18*, 2519–2536. [CrossRef]
22. Fursova, A.Z.; Derbeneva, A.S.; Tarasov, M.S.; Nikulich, I.F.; Devyatkin, V.A.; Telegina, D.V.; Kolosova, N.G.; Kozhevnikova, O.S. Leukocyte telomere length and response to antiangiogenic therapy in patients with neovascular age-related macular degeneration. *Adv. Gerontol.* **2022**, *12*, 135–142. [CrossRef]
23. Solé, X.; Guinó, E.; Valls, J.; Iniesta, R.; Moreno, V. SNPStats: A web tool for the analysis of association studies. *Bioinformatics* **2006**, *22*, 1928–1929. [CrossRef]
24. Kaarniranta, K.; Blasiak, J.; Liton, P.; Boulton, M.; Klionsky, D.J.; Sinha, D. Autophagy in age-related macular degeneration. *Autophagy* **2023**, *19*, 388–400. [CrossRef] [PubMed]
25. Wang, Y.; Yao, Y.; Li, R.; Wu, B.; Lu, H.; Cheng, J.; Liu, Z.; Du, J. Different effects of anti-VEGF drugs (Ranibizumab, Aflibercept, Conbercept) on autophagy and its effect on neovascularization in RF/6A cells. *Microvasc. Res.* **2021**, *138*, 104207. [CrossRef] [PubMed]
26. Hu, Y.L.; Jahangiri, A.; De Lay, M.; Aghi, M.K. Hypoxia-induced tumor cell autophagy mediates resistance to anti-angiogenic therapy. *Autophagy* **2012**, *8*, 979–981. [CrossRef] [PubMed]
27. Wang, Y.; Fung, N.S.K.; Lam, W.C.; Lo, A.C.Y. mTOR Signalling Pathway: A Potential Therapeutic Target for Ocular Neurodegenerative Diseases. *Antioxidants* **2022**, *11*, 1304. [CrossRef]
28. Fernandes, S.A.; Demetriades, C. The Multifaceted Role of Nutrient Sensing and mTORC1 Signaling in Physiology and Aging. *Front. Aging* **2021**, *2*, 707372. [CrossRef]
29. Merle, D.A.; Provenzano, F.; Jarboui, M.A.; Kilger, E.; Clark, S.J.; Deleidi, M.; Armento, A.; Ueffing, M. mTOR Inhibition via Rapamycin Treatment Partially Reverts the Deficit in Energy Metabolism Caused by FH Loss in RPE Cells. *Antioxidants* **2021**, *10*, 1944. [CrossRef]
30. Tamargo-Gómez, I.; Fernández, Á.F.; Mariño, G. Pathogenic Single Nucleotide Polymorphisms on Autophagy-Related Genes. *Int. J. Mol. Sci.* **2020**, *21*, 8196. [CrossRef]

31. Nakatogawa, H. Mechanisms governing autophagosome biogenesis. *Nat. Rev. Mol. Cell Biol.* **2020**, *21*, 439–458.
32. Zachari, M.; Ganley, I.G. The Mammalian ULK1 Complex and Autophagy Initiation. *Essays Biochem.* **2017**, *61*, 585–596. [CrossRef] [PubMed]
33. Henckaerts, L.; Cleynen, I.; Brinar, M.; John, J.M.; Van Steen, K.; Rutgeerts, P.; Vermeire, S. Genetic variation in the autophagy gene ULK1 and risk of Crohn's disease. *Inflamm. Bowel Dis.* **2011**, *17*, 1392–1397. [PubMed]
34. Zhang, R.R.; Liang, L.; Chen, W.W.; Wen, C.; Wan, B.S.; Luo, L.L.; Zhao, Y.L.; Chen, J.; Yue, J. ULK1 polymorphisms confer susceptibility to pulmonary tuberculosis in a Chinese population. *Int. J. Tuberc. Lung Dis.* **2019**, *23*, 265–271. [CrossRef] [PubMed]
35. Zhang, X.; Han, R.; Wang, M.; Li, X.; Yang, X.; Xia, Q.; Liu, R.; Yuan, Y.; Hu, X.; Chen, M.; et al. Association between the autophagy-related gene ULK1 and ankylosing spondylitis susceptibility in the Chinese Han population: A case-control study. *Postgrad. Med. J.* **2017**, *93*, 752–757.
36. Katsuragi, Y.; Ichimura, Y.; Komatsu, M. P62/SQSTM1 Functions as a Signaling Hub and an Autophagy Adaptor. *FEBS J.* **2015**, *82*, 4672–4678. [CrossRef]
37. Johansen, T.; Lamark, T. Selective Autophagy: ATG8 Family Proteins, LIR Motifs and Cargo Receptors. *J. Mol. Biol.* **2020**, *432*, 80–103. [CrossRef]
38. Haack, T.B.; Ignatius, E.; Calvo-Garrido, J.; Iuso, A.; Isohanni, P.; Maffezzini, C.; Lönnqvist, T.; Suomalainen, A.; Gorza, M.; Kremer, L.S.; et al. Absence of the Autophagy Adaptor SQSTM1/p62 Causes Childhood-Onset Neurodegeneration with Ataxia, Dystonia, and Gaze Palsy. *Am. J. Hum. Genet.* **2016**, *99*, 735–743. [CrossRef]
39. Teyssou, E.; Takeda, T.; Lebon, V.; Boillée, S.; Doukouré, B.; Bataillon, G.; Sazdovitch, V.; Cazeneuve, C.; Meininger, V.; Leguern, E.; et al. Mutations in SQSTM1 encoding p62 in amyotrophic lateral sclerosis: Genetics and neuropathology. *Acta Neuropathol.* **2013**, *125*, 511–522. [CrossRef]
40. Le Ber, I.; Camuzat, A.; Guerreiro, R.; Bouya-Ahmed, K.; Bras, J.; Nicolas, G.; Gabelle, A.; Didic, M.; De Septenville, A.; Millecamps, S.; et al. SQSTM1 Mutations in french patients with frontotemporal dementia or frontotemporal dementia with amyotrophic lateral sclerosis. *JAMA Neurol.* **2013**, *70*, 1403–1410.
41. Fursova, A.Z.; Nikulich, I.F.; Dmitrieva, E.I.; Gusarevich, O.G.; Derbeneva, A.S.; Vasilyeva, M.A.; Kozhevnikova, O.S.; Kolosova, N.G. Three-year follow-up study of clinical effectiveness of antiangiogenic therapy for neovascular age-related macular degeneration. *Vestn. Oftalmol.* **2023**, *139*, 45–52. [CrossRef] [PubMed]

MDPI
St. Alban-Anlage 66
4052 Basel
Switzerland
www.mdpi.com

Biomedicines Editorial Office
E-mail: biomedicines@mdpi.com
www.mdpi.com/journal/biomedicines